SpringerBriefs in Computer Science

For further volumes:
http://www.springer.com/series/10028

John F. Dooley

A Brief History of Cryptology and Cryptographic Algorithms

John F. Dooley
Department of Computer Science
Knox College
Galesburg, IL
USA

ISSN 2191-5768 ISSN 2191-5776 (electronic)
ISBN 978-3-319-01627-6 ISBN 978-3-319-01628-3 (eBook)
DOI 10.1007/978-3-319-01628-3
Springer Cham Heidelberg New York Dordrecht London

Library of Congress Control Number: 2013945798

Printed on acid-free paper

Springer is part of Springer Science+Business Media (www.springer.com)

For diane

Preface

Cryptology is the science of secret communications. You are likely to use some form of cryptology every day. If you login to a computer you are using cryptology in the form of a one-way hash function that protects your password. If you buy something over the Internet, you are using two different forms of cryptology—public-key cryptography to set up the encrypted network connection between you and the vendor and a symmetric key algorithm to finish your transaction. These days much of the cryptology that is in use is invisible, just like the examples given. It wasn't always so. The story of cryptology goes back at least 2,500 years and for most of that time it was considered an arcane science, known only to a few and jealously guarded by governments, exiled kings and queens, and religious orders. For a time in the European Middle Ages it was even considered to be a form of magic. It is only recently, really beginning in the twentieth century, that cryptology has become known and studied outside the realms of secret government agencies. Even more recently, the study of cryptology has moved from a branch of linguistics to having a firm foundation in mathematics.

This book is a brief history of cryptology from the time of Julius Caesar up through around the year 2001. It also covers the different types of cryptographic algorithms used to create secret messages, and it discusses methods for breaking secret messages. There are several examples in the text that illustrate the algorithms in use. Being 'brief', it is not meant to be a comprehensive history of either cryptology or the algorithms themselves. Rather I have tried to touch on a subset of the important stories in cryptologic history and the algorithms and people involved. Most of the chapters begin with a story that tries to illustrate the importance of cryptology in that particular time period.

I teach an upper-level undergraduate survey course in *Cryptography and Computer Security* and the contents of this book are about the first quarter of that course where I do a review of the different cryptographic algorithms from a historical perspective. My goal in that part of the course is to give students a better understanding of *how* we got from the early days of pencil and paper secret messages to a place where cryptology is pervasive and invisible. This book could easily serve as the text for that part of a course on computer or network security or as a supplemental text for a stand-alone course on computer security. No mathematics is required beyond what a computer science or mathematics student would see in a course on discrete mathematics. If you want to pursue a more comprehensive treatment of the history of cryptology, I recommend David Kahn's excellent book *The Codebreakers: The Story of Secret Writing,* and for a more mathematical treatment, Craig Bauer's equally good *Secret History: The Story of Cryptology*.

Acknowledgments

I would like to thank the library staff at Knox College for their patience and professional help in finding copies of many of the articles and letters referenced here. I would also like to thank the staff of the National Archives and Records Administration (NARA) in College Park, MD, Librarian René Stein at the Research Library at the National Cryptologic Museum in Ft. Meade, MD, and Paul Barron and Jeffrey Kozak of the George C. Marshall Foundation Research Library in Lexington, VA for their excellent help. And of course, thanks to Diane, who inspires me, encourages me, and—above the call of duty—reads and edits everything I write.

This research was funded in part by a grant from the Andrew W. Mellon Foundation and by the Office of the Dean of the College at Knox College.

Contents

Chapter 1
Introduction: A Revolutionary Cipher

Abstract Cryptology is the science of secret writing. It is made up of two halves; cryptography consists of the techniques for creating systems of secret writing and cryptanalysis encompasses the techniques of breaking them. Over the past 2,500 years, cryptology has developed numerous types of systems to hide messages and subsequently a rich vocabulary in which to describe them. In this chapter we introduce the reader to the vocabulary of cryptology, explain the differences between codes and ciphers and begin the discussion of how to decipher an unknown message.

1.1 A Traitorous Doctor

In the summer of 1775, the American revolutionary forces were near a state of chaos. The main body of the American force was laying siege to Boston. The Continental Congress had just appointed George Washington of Virginia as commander of all continental forces. Money was scarce, enlistments were short, and most of the Continental Army was comprised of colonial militias with little training, no common equipment, and no idea of the enemy they faced. The officer corps was not in much better shape, with most of the colonial officers having had little or no command experience. Logistics were haphazard, artillery was practically non-existent, and the British held all the major urban areas in the thirteen colonies. The last thing that Lieutenant General Washington needed in September 1775 was a Tory spy in his midst sending secret messages to the British. But that is exactly what he got.

In mid-August 1775 a young patriot from Newport, Rhode Island named Godfrey Wenwood received a request from a former lover. It was to deliver a letter to a "Major Cane in Boston on his magisty's service". Wenwood was rather reluctant to deliver the letter, assuming, quite correctly, that Major Cane was a British

J. F. Dooley, *A Brief History of Cryptology and Cryptographic Algorithms*, SpringerBriefs in Computer Science, DOI: 10.1007/978-3-319-01628-3_1,

officer stationed in Boston with access to General Gage, the commander of British forces in America. Instead he took it to a friend of his, a fellow patriot and a schoolmaster, who opened it and discovered three sheets of unintelligible writing. The friend could not decipher the message and gave it back to Wenwood, who proceeded to sit on the letter for nearly two months. Figure 1.1 shows a page from the letter. Only when prompted by another letter from his former lover (whose name and fate have been lost to history) asking why the first one had yet to be delivered

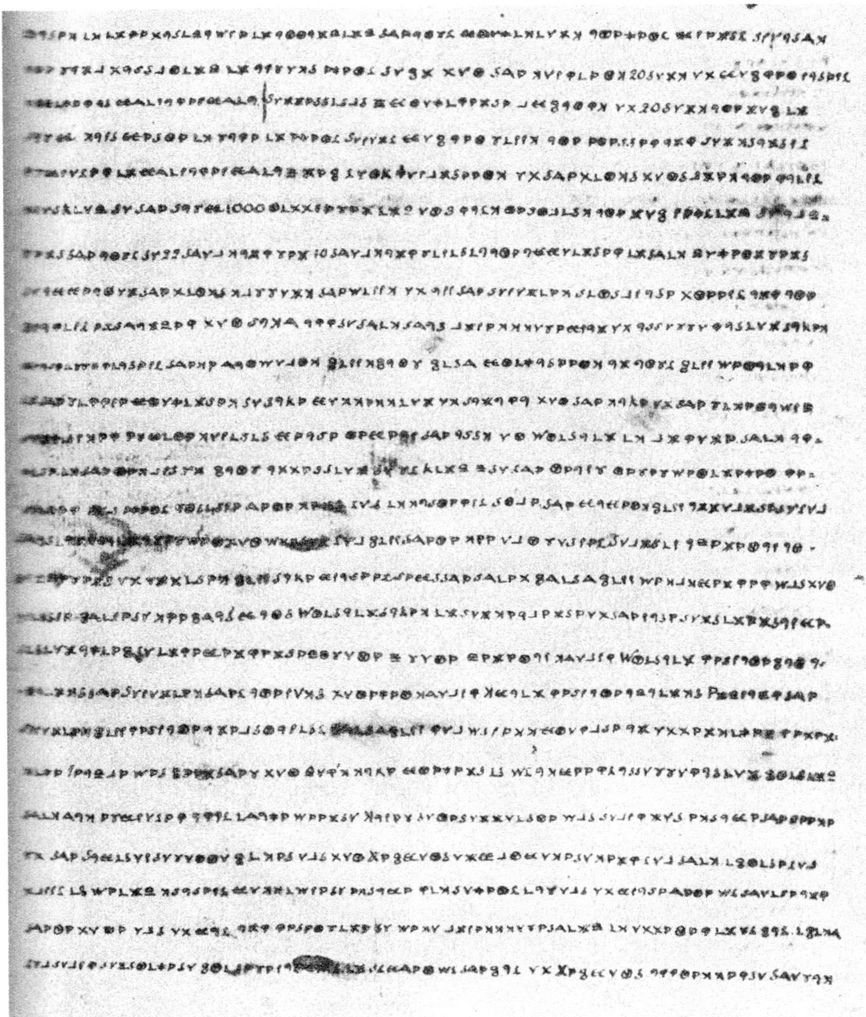

Fig. 1.1 Page from Dr. Church's cipher letter (Lib of Congress)

did Wenwood act. At the end of September 1775, he traveled the sixty-five miles from Newport to Washington's headquarters in Cambridge, Massachusetts and delivered the letter in person to General Washington.

Of course Washington, who couldn't read the letter either, ordered the woman arrested and brought to his camp for questioning. At the end of a lengthy interrogation—performed mostly by Washington himself—she gave up the name of the author of the letter—Dr. Benjamin Church, Jr., her current lover.

Dr. Church was a seemingly devoted revolutionary, a member of the Massachusetts Provincial Congress, and the head of the nascent army's medical corps as Washington's director general of hospitals. A well-to-do Boston physician, and a Harvard graduate, he was a friend of John Hancock and Samuel Adams. Dr. Church ran in all the best revolutionary circles. He was also a sham—a Loyalist to the core who had been a British spy since at least 1774, regularly reporting first to the Governor of Massachusetts and then to General Gage.

Church was brought in for questioning, and immediately acknowledged authorship of the letter. He said, despite the address on the outside, that the letter was intended for his brother in Boston and that the contents were entirely innocuous. But he refused to decipher the letter for Washington.

Washington still couldn't read the now very suspicious letter, but he thought he might know people who could. In the eighteenth century, because letters were mailed just by folding the paper on which they were written and sealing with wax, many people enciphered ordinary mail to maintain their own privacy. So there were officers in the continental army who had some familiarity with ciphers. Washington gave copies of the letter to two people, the Reverend Samuel West, a Massachusetts militia chaplain, and Elbridge Gerry, future Vice-President of the United States and originator of the gerrymander. Gerry also recruited Colonel Elisha Porter of the Massachusetts militia to help. With Gerry and Porter together, and West alone, the two teams, worked through the night, producing two identical solutions. This was the first successful cryptanalysis of the American Revolution. The letter was written in a simple monoalphabetic substitution cipher and was a blockbuster [1, pp. 541–542].

The contents of the letter were not quite damning. While Church gave much information about American army strengths and weaknesses, the letter also seemed to convey the determination of the colonists in the fight for freedom. The most damaging parts are where Church is describing how to send him correspondence—"I wish you could contrive to write me largely in cipher, by the way of Newport, addressed to Thomas Richards, Merchant." And the last line of the letter, that convinced Washington and his officers that Church was a Tory spy—"Make use of every precaution or I perish."

Washington had Church imprisoned while awaiting formal charges and a trial; a trial that never came. In 1777 the British offered to exchange Church for a captured American surgeon, but Congress declined. Finally, in 1780 Congress ordered Church exiled to the West Indies. He was put on a schooner, which sailed from Boston and was never heard of again, apparently lost at sea [2, pp. 174–176].

1.2 A Few (Vocabulary) Words About Cryptology

Secret writing is known to have existed for close to 2,500 years. As Kahn puts it, "It must be that as soon as a culture has reached a certain level, probably measured largely by its literacy, cryptography appears spontaneously—as its parents, language and writing, probably also did. The multiple human needs and desires that demand privacy among two or more people in the midst of social life must inevitably lead to cryptology wherever men thrive and wherever they write. Cultural diffusion seems a less likely explanation for its occurrence in so many areas, many of them distant and isolated." [2, p. 84]

Every discipline has its own vocabulary and cryptology is no different. This section does not attempt to be a comprehensive glossary of cryptology, but rather gives the basic definitions and jargon. Many of the concepts introduced here will be explored further in the chapters to come.

Cryptology is the study of secret writing. Governments, the military, and people in business have desired to keep their communications secret ever since the invention of writing. Spies, lovers, and diplomats all have secrets and are desperate to keep them as such. There are typically two ways of keeping secrets in communications. *Steganography* hides the very existence of the message. Secret ink, microdots, and using different fonts on printed pages are all ways of hiding the message from prying eyes. *Cryptology*, on the other hand, makes absolutely no effort to hide the presence of the secret message. Instead it transforms the message into something unintelligible so that if the enemy intercepts the message they will have no hope of reading it. A *cryptologic system* performs a *transformation* on a message—called the *plaintext*. The transformation renders the plaintext unintelligible and produces a new version of the message—the *ciphertext*. This process is *encoding* or *enciphering* the plaintext. A message in ciphertext is typically called a *cryptogram*. To reverse the process the system performs an inverse transformation to recover the plaintext. This is known as *decoding* or *decrypting* the ciphertext.

The science of cryptology can be broken down in a couple of different ways. One way to look at cryptology is that it is concerned with both the creation of cryptologic systems, called *cryptography* and with techniques to uncover the secret from the ciphertext, called *cryptanalysis*. A person who attempts to break cryptograms is a *cryptanalyst*. A complementary way of looking at cryptology is to divide things up by the types and sizes of grammatical elements used by the transformations that different cryptologic systems perform. The standard division is by the size of the element of the plaintext used in the transformation. A *code* uses variable sized elements that have meaning in the plaintext language, like syllables, words, or phrases. On the other hand, a *cipher* uses fixed sized elements like single letters or two- or three-letter groups that are divorced from meaning in the language. For example, a code will have a single *codeword* for the plaintext "stop", say 37761, while a cipher will transform each individual letter as in X = s, A = t, V = o, and W = p to produce XAVW. One could argue that a code is

Table 1.1 The two dimensions of Cryptology

	Cryptography		Cryptanalysis			
Codes	1-part	2-part	Theft, spying	Probable word	Context	
Ciphers	Substitution	Transposition	Classical	Statistical	Mathematical	Brute-force
	Product cipher					

also a substitution cipher, just one with a larger number of substitutions. However, while ciphers have a small fixed number of substitution elements—the letters of the alphabet—codes typically have thousands of words and phrases to substitute. Additionally, the methods of cryptanalysis of the two types of system are quite different.

Table 1.1 provides a visual representation of the different dimensions of cryptology.

1.3 Codes

A *code* always takes the form of a book where a numerical or alphabetic *codeword* is substituted for a complete word or phrase from the plaintext. *Codebooks* can have thousands of codewords in them. There are two types of codes, 1-part and 2-part. In a 1-part code there is a single pair of columns used for both encoding and decoding plaintext. The columns are usually sorted so that lower numbered codewords will correspond to plaintext words or phrases that are lower in the alphabetic ordering. For example,

1234	Centenary
1235	Centennial
1236	Centime
1237	Centimeter
1238	Central nervous system

Note that because both the codewords and the words they represent are in ascending order, the *cryptanalyst* will instantly know that a codeword of 0823 must begin with an alphabetic sequence before "ce", thus eliminating many possible codeword-plaintext pairs.

A 2-part code eliminates this problem by having two separate lists, one arranged numerically by codewords and one arranged alphabetically by the words and phrases the codewords represent. Thus one list (the one that is alphabetically sorted) is used for encoding a message and the other list (the one that is numerically sorted by codeword) is used for decoding messages. For example, the list used for encoding might contain

Artillery support	18312
Attack	43110
Company	13927
Headquarters	71349
Platoon strength	63415

while the decoding list would have

13927	Company
18312	Artillery support
43110	Attack
63415	Platoon strength
71349	Headquarters

Note that not only are the lists not compiled either numerically or alphabetically, but also there are gaps in the list of codewords to further confuse the cryptanalyst.

Cryptanalyzing codes is very difficult because there is no logical connection between a codeword and the plaintext code or phrase it represents. With a 2-part code there is normally no sequence of codewords that represent a similar alphabetical sequence of plaintext words. Because a code will likely have thousands of codeword-plaintext pairs, the cryptanalyst must slowly uncover each pair and over time create a dictionary that represents the code. The correspondents may make this job easier by using standard salutations or formulaic passages like "Nothing to report" or "Weather report from ship AD2342". If the cryptanalyst has access to enough ciphertext messages then sequences like this can allow her to uncover plaintext. Still, this is a time-consuming endeavor. Of course the best way to break a code is to steal the codebook! As we will see, this has happened a number of times in history, much to the dismay of the owner.

Codes have issues for users as well. Foremost among them is distributing all the codebooks to everyone who will be using the code. Everyone who uses a code must have exactly the same codebook and must use it in exactly the same way. This limits the usefulness of codes because the codebook must be available whenever a message needs to be encoded or decoded. The codebook must also be kept physically secure, ideally locked up when not in use. If one copy of a codebook is lost or stolen, then the code can no longer be used and every copy of the codebook must be replaced. This makes it hard to give codebooks to spies who are traveling in enemy territory, and it also makes it very difficult to use codes in battlefield situations where they could be easily lost.

1.4 Ciphers

This brings us to *ciphers*. Ciphers also transform plaintext into ciphertext, but unlike codes, ciphers use small, fixed-length language elements that are divorced from the meaning of the word or phrase in the message. Ciphers come in two general categories. *Substitution ciphers* will replace each letter in a message with a different letter or symbol using a mapping called a *cipher alphabet*. The second type will rearrange the letters of a message, but will not substitute new letters for the existing letters in the message. These are *transposition ciphers*.

1.5 Substitution Ciphers

Substitution ciphers can use just a single cipher alphabet for the entire message; these are known as *monoalphabetic substitution ciphers*. Cipher systems that use more than one cipher alphabet to do the encryption are *polyalphabetic substitution ciphers*. In a polyalphabetic substitution cipher each plaintext letter may be replaced with more than one *cipher letter*, making the job significantly harder for the cryptanalyst. The cipher alphabets may be *standard alphabets* that are shifted using a simple key. For example a shift of 7 results in,

```
Plain:  abcdefghijklmnopqrstuvwxyz
Cipher: HIJKLMNOPQRSTUVWXYZABCDEFG
```

And the word *attack* becomes HAAHJR. Or they may be *mixed alphabets* that are created by a random rearrangement of the standard alphabet as in

```
Plain:  abcdefghijklmnopqrstuvwxyz
Cipher: BDOENUZIWLYVJKHMFPTCRXAQSG
```

And the word *enemy* is transformed into NKNJS.

All substitution ciphers depend on the use of a *key* to tell the user how to rearrange the standard alphabet into a cipher alphabet. If the same key is used to both encrypt and decrypt messages then the system is called a *symmetric key system*.

Just like the security of a codebook, the security of the key is of paramount importance for cipher systems. And just like a codebook, everyone who uses a particular cipher system must also use the same key. For added security, keys are changed periodically, so while the basic substitution cipher system remains the same, the key is different. Distributing new keys to all the users of a cryptologic system leads to the *key management problem*. Management of the keys is a problem because a secure method must be used to transmit the keys to all users. Typically, a courier distributes a book listing all the keys for a specific time period, say a month, and each user has instructions on when and how to change keys. And just like codebooks, any loss or compromise of the key book will jeopardize the system. But unlike codebooks, if a key is lost the underlying cipher system is not compromised and merely changing the key will restore the integrity of the cipher system.

While most cipher systems substitute one letter at a time, it is also possible to substitute two letters at a time, called a *digraphic* system, or more than two, called a *polygraphic* system. A substitution cipher that provides multiple substitutions for some letters but not others is a *homophonic* system. It is also possible to avoid the use of a specific cipher alphabet and use a book to identify either individual letters or words. This is known as a *book* or *dictionary cipher*. The sender specifies a particular page, column, and word in the book for each word or letter in the plaintext and the recipient looks up the corresponding numbers to decrypt the message. For example, a codeword of 0450233 could specify page 045, column 02, and word 33 in that column. Naturally, the sender and recipient must each have a copy of exactly the same edition of the book in order for this system to work. But carrying a published book or dictionary is significantly less suspicious than a codebook.

1.6 Transposition Ciphers

Transposition ciphers transform the plaintext into ciphertext by rearranging the letters of the plaintext according to a specific rule and key. The transposition is a *permutation* of all the letters of the plaintext message done according to a set of rules and guided by the key. Since the transposition is a permutation, there are n! different cipher texts for an n-letter plaintext message. The simplest transposition cipher is the *columnar transposition*. This comes in two forms, the *complete columnar transposition* and the *incomplete columnar*. In both of these systems, the plaintext is written horizontally in a rectangle that is as wide as the length of the key. As many rows as are needed to complete the message are used. In the complete columnar transposition once the plaintext is written out the columns are then filled with nulls until they are all the same length. For example,

```
s e c o n d
d i v i s o
n a d v a n
c i n g t o
n i g h t x
```

The ciphertext is then pulled off by columns according to the key and divided into groups of five for transmission. If the key for this cipher were 321654 then the ciphertext would be

```
cvdng eiaii sdncn donox nsatt oivgh
```

An *incomplete columnar transposition cipher* doesn't require complete columns and so leaves off the null characters resulting in columns of differing lengths and making the system harder to cryptanalyze. Another type of columnar transposition cipher is the *route transposition*. In a route transposition, one creates the standard rectangle of the plaintext, but then one takes off the letters using a rule that describes a route through the rectangle. For example, one could start at the upper left-hand corner and describe a spiral through the plaintext, going down one column, across a row, up a column and then back across another row. Another method is to take the message off by columns, but alternate going down and up each column.

Cryptanalysis of ciphers falls into four different, but related areas. The *classical* methods of cryptanalysis rely primarily on language analysis. The first thing the cryptanalyst must know about a cryptogram is the language in which it is written. Knowing the language is crucial because different languages have different language characteristics, notably letter and word frequencies and sentence structure. It turns out that if you look at several pieces of text that are several hundred words long and written in the same language that the frequencies of all the letters used turn out to be about the same in all of the texts. In English, the letter 'e' is used about 13 % of the time, 't' is used about 10 % of the time, etc. down to 'z', which is used less than 1 % of the time. So the cryptanalyst can count each of the letters in a cryptogram and get a hint of what the substitutions may have been.

Beginning in the early 20th century, cryptanalysts began applying *statistical* tests to messages in an effort to discern patterns in more complicated cipher systems, particularly in polyalphabetic systems. Later in the 20th century, with the introduction of machine cipher systems, cryptanalysts began applying more *mathematical analysis* to the systems, particularly bringing to bear techniques from combinatorics, algebra, and number theory. And finally, with the advent of computers and computer cipher systems in the late 20th century, cryptanalysts have had to fall back on *brute-force* guessing to extract the key from a cryptogram or, more likely, a large set of cryptograms.

References

1. Freeman, Douglas Southall. 1951. *George Washington: Planter and patriot*. New York: Charles Scribner's Sons.
2. Kahn, David. 1967. *The codebreakers: The story of secret writing*. New York: Macmillan.

Chapter 2
Cryptology Before 1500: A Bit of Magic

Abstract Cryptology was well established in ancient times, with both Greeks and Romans practicing different forms of cryptography. With the fall of the Roman Empire, cryptology was lost in the West until the Renaissance, but it flourished in the Arabic world. The Arabs invented the first reliable tool for cryptanalysis, frequency analysis. With the end of the Middle Ages and the increase in commerce and diplomacy, cryptology enjoyed a Renaissance of its own in the West. This chapter examines the most common cipher of the period, the monoalphabetic substitution cipher and then looks at the technique of frequency analysis that is used to break the monoalphabetic substitution. An extended example is given to illustrate the use of frequency analysis to break a monoalphabetic.

2.1 Veni, Vidi, Cipher

Julius Caesar, probably the greatest of all Roman generals, was no stranger to cryptology. In his famous *Commentary on the Gallic Wars*, Caesar himself describes using a form of a cipher to hide a message.

> Then with great rewards he induces a certain man of the Gallic horse to convey a letter to Cicero. *This he sends written in Greek characters, lest the letter being intercepted, our measures should be discovered by the enemy.* He directs him, if he should be unable to enter, to throw his spear with the letter fastened to the thong, inside the fortifications of the camp. He writes in the letter, that he having set out with his legions, will quickly be there: he entreats him to maintain his ancient valor. The Gaul apprehending danger, throws his spear as he has been directed. It by chance stuck in a tower, and, not being observed by our men for two days, was seen by a certain soldier on the third day: when taken down, it was carried to Cicero. He, after perusing it, reads it out in an assembly of the soldiers, and fills all with the greatest joy. Then the smoke of the fires was seen in the distance, a circumstance which banished all doubt of the arrival of the legions [1, Chap. 48, italics added].

This, however, is not Caesar's most famous contribution to the history of cryptology. The Roman historian Gaius Suetonius Tranquillus, in his *The Twelve Caesars* describes Julius Caesar's use of a cipher to send messages to his friends

J. F. Dooley, *A Brief History of Cryptology and Cryptographic Algorithms*,
SpringerBriefs in Computer Science, DOI: 10.1007/978-3-319-01628-3_2,

and political allies. This was a cipher that, according to Seutonius, "If he had anything confidential to say, he wrote it in cipher, that is, by so changing the order of the letters of the alphabet, that not a word could be made out. If anyone wishes to decipher these, and get at their meaning, he must substitute the fourth letter of the alphabet, namely D, for A, and so with the others" [3, Chap. 56]. This is the first written description of the modern monoalphabetic substitution cipher using a shifted standard alphabet. Using Caesar's cipher, the cipher alphabet looks like

```
Plain:  abcdefghijklmnopqrstuvwxyz
Cipher: DEFGHIJKLMNOPQRSTUVWXYZABC
```

and Caesar's famous "I came, I saw, I conquered" would be enciphered as `L FDPH, L VDZ, L FRQTXHUHG.`

2.2 Cryptology in the Middle Ages

For 900 years the monoalphabetic substitution cipher was the strongest cipher system in the Western world. The Romans used it regularly to protect their far-flung lines of communication. But after the fall of the Western Roman Empire in 476 C.E. the knowledge of cryptology vanished from the West and wasn't to return until the Italian Renaissance. Indeed, with the decline of literacy and scholarship in Europe during the Dark Ages following the fall of Rome cryptology turned from a useful technique for keeping communications secret into a dark art that bordered on magic.

But interest in cryptology was not dead. In the latter part of the first millennium, there was another place where intellectual curiosity and scholarship flowered and where mathematics and cryptology saw their biggest advances since Caesar—the Arab world. And this was where the next big advance in cryptanalytic techniques would come from.

The period around the 9th century C.E. is considered to be the beginning of the Islamic Golden Age, when philosophy, science, literature, mathematics, and religious studies all flourished in what was then the peace and prosperity of the Abbasid Caliphate. Into this period was born Abu Yūsuf Ya-qūb ibn Isāq as-Sabbāh al-Kindi (801–873 C.E.), a polymath who was the philosopher of the age. Al-Kindi wrote books in many disciplines including astronomy, optics, philosophy, mathematics, medicine, and linguistics, but his book on secret messages for court secretaries, A *Manuscript on Deciphering Cryptographic Messages* is the most important to the history of cryptology. It is in this book that the technique of *frequency analysis* is first described.

2.3 Frequency Analysis, the First Cryptanalytic Tool

In every language, if one is given a text of several hundred or thousand characters and the individual letters in the text are counted, some of the letters will appear more often than others, and some will appear very infrequently. If another text of

Table 2.1 English frequency percentages

Letter	Percentage	Letter	Percentage
A	8.4	N	7.0
B	1.7	O	7.3
C	3.1	P	2.0
D	4.4	Q	0.1
E	12.7	R	6.3
F	2.0	S	6.0
G	2.0	T	9.3
H	5.4	U	2.4
I	7.0	V	1.0
J	0.2	W	2.0
K	0.7	X	0.2
L	4.0	Y	2.2
M	2.5	Z	0.1

similar length is analyzed in the same way, the same letters will pop up as either more frequently occurring or less frequently occurring. Thus, the *frequency of occurrence* of individual letters is a characteristic of the language.

It is also impossible to hide this frequency of occurrence if one substitutes one letter for another in a message. What al-Kindi discovered is that in a message enciphered using a monoalphabetic substitution cipher, the language characteristics are not hidden by the substitution. In particular the letter frequencies will shine through the substitution like a beacon leading the cryptanalyst to the concealed letters of the plaintext.

In English, the most frequently occurring letters are usually given in the order of ETAOINSHRDLU. Table 2.1, which was constructed by counting all 95,512 or so words (450,583 letters) in David Kahn's biography of Herbert O. Yardley, *The Reader of Gentlemen's Mail* illustrates the ordering for modern English usage.

Graphically, this looks like Fig. 2.1.

The technique of frequency analysis is to do the same count of letters for the ciphertext, and then use those counts to guess at the letters of the ciphertext. Thus, the most frequently occurring letter in the ciphertext should represent e. The next most frequently occurring should represent t, then a, etc. al-Kindi laid all this out in a few short paragraphs and with it revolutionized cryptanalysis.

One does not need to be restricted to just single letter frequencies when doing this type of analysis. It turns out that there are also pairs of letters (digraphs) that occur with great frequency and pairs that don't occur at all. For example, in English, the most frequent pairs of letters are *th*, *he*, *in*, *er*, *an*, *re*, and *nd*. And one could continue with the most common three letter words in English, *the*, *and*, *for*, *not*, and *you*.

To illustrate the technique of frequency analysis, lets decrypt an English cryptogram that was created using a monoalphabetic substitution cipher. How should we go about decrypting the following cryptogram?

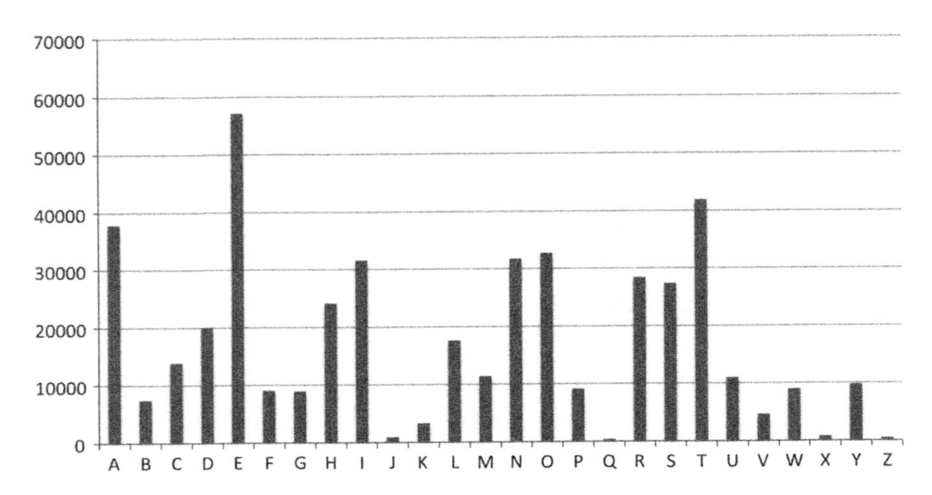

Fig. 2.1 A graph of English letter frequencies

```
SCEAC SKDXA CESDS CKVSO LCDDA GKEMG AMTYK TOVKS OSFNC
FPCEE XMTDA OLTCQ OLGKG ACOKS ADSFN EGFGN KCHLQ HGFOL
TMQRI TYOSF VLSYL SCFCD XMTGF TLQFP KTPCF PMSWO XMTHC
KCOTY SHLTK MRQOS YGFAT MMOLC OOLSM SMTFO SKTDX FTVOG
ETOLT GRITY OGAOL GMTVL GSFUT FOTPO LTMXM OTELC MCHHC
KTFOD XRTTF OGYGF YTCDO LCOOL TMTYL CKCYO TKMYG FUTXC
ETMMC NTCFP OGNSU TOLTS PTCOL COOLT XCKTO LTETK TKCFP
GEMBT OYLTM GAYLS DPKTF CKOLQ KYGFC FPGXD TOLTC PUTFO
QKTGA OLTPC FYSFN ETF
```

We begin by counting all the letters in the cryptogram and producing two things—a frequency table and a frequency chart. The frequency table looks like Table 2.2.

And the frequency chart for the cryptogram looks like Fig. 2.2.

Looking at the many ups and downs in the frequency chart we can easily see that this is a monoalphabetic substitution. With the T being so much higher than any of the other letters, it is our top candidate for *e*. O and C look like candidates to be the next two highest frequency letters *t* and *a*, but which is which we don't know yet. Remember that the frequency count for English is based on a very large number of letters, while the frequency count for a single cryptogram is based on many fewer letters. That fact may skew some of the frequencies and the overall distribution.

Our next step is to try to break down the letters in the cryptogram into at least three different groups—high frequency letters, medium frequency, and low frequency. In standard English, *e, t, a, i, o, n, r, s,* and *h* form the high-frequency letters—defined as those with a frequency percentage of greater than 5 % for our purposes, For the medium frequency group we have *c, d, f, g, l, m, p, u, w*, and *y* and for the low-frequency letters (at less than 2 % of the count each) we have *b, j, k, q, v, x,* and *z*. So if we can identify these groups in the cryptogram we could be

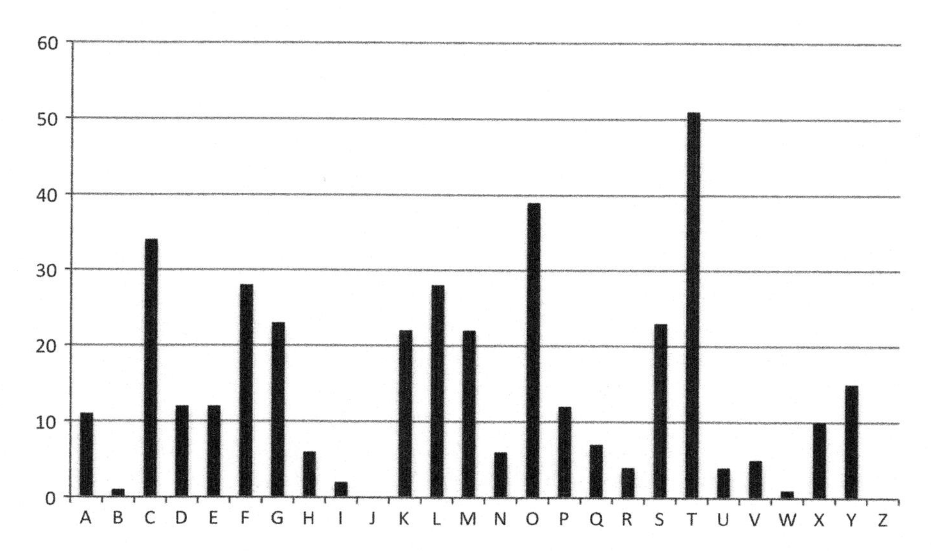

Fig. 2.2 Cryptogram frequency chart

Table 2.2 Cryptogram frequency count

A	B	C	D	E	F	G	H	I	J	K	L	M
11	1	34	12	12	28	23	6	2	0	22	28	22
N	O	P	Q	R	S	T	U	V	W	X	Y	Z
6	39	12	7	4	23	51	4	5	1	10	15	0

Table 2.3 Frequency count in descending order

T	O	C	F	L	G	S	K	M	Y	D	E	P
51	39	34	28	28	23	23	22	22	15	12	12	12
A	X	Q	H	N	V	R	U	I	B	W	J	Z
11	10	7	6	6	5	4	4	2	1	1	0	0

on our way to getting the entire cipher alphabet. If we re-arrange Table 2.2 so that the letters are written in descending order by count, we get Table 2.3.

Ignoring the large dip between the T and the O, the next big dip in frequency is a dip of 7 between the M and the Y, conveniently between the ninth and tenth letters, just where the dip between the high and medium frequency letters is. Now that we have a feel for how the individual letters are arranged, it is time to look at digraphs. Digraphs give us a feel for how the letters arrange themselves next to other letters. We've seen that *th*, *he*, *in*, *er*, *an*, *re*, and *nd* are the most common digraphs, so it should be the case that some pairs of letters in the cryptogram behave similarly.

Looking at the digraphs we see that OL is the most frequently occurring digraph at 18. LT occurs 12 times (and the three-letter group OLT occurs 9 times),

KT eight times, MT, CF, GF, and TF all occur seven times, and TM occurs six times. If we assume that OL is the digraph *th*, and LT is the digraph *he*, we then have good confirmation that O = t, L = h, and T = e.

The next thing is to identify the other high-frequency letters, especially the vowels, *a*, *i*, and *o*. The next three highest frequency ciphertext letters are C, F, and L. We also note that the sequence OLCO occurs three times in the cryptogram. Given what we already know, this sequence decrypts to *th*t*, which could be the word *that*, leaving C = a. This replacement also gives us the popular digraph *ea* five times in the deciphered part of the cryptogram, a good sign.

The next high frequency digraph is *in* which also includes two letters from the high-frequency letter group. Looking carefully through the ciphertext, we see that S occurs 23 times and F occurs 28 times. This might lead us to believe that F = *i* and S = *n*. If we substitute these new pairs, however, we get decrypted sequences like LSCFC = *hnaia* and OLCOOLS = *thatths*, neither of which look promising. If instead we see that the digraph SF occurs 5 times and the trigraph SFN occurs twice we can go further. If SF = *in* then it is possible that SFN = *ing* allowing us to supposed that S = *i*, F = *n*, and N = *g*. This will also give us the trigraph *ent* in 5 different places; another good sign. Putting those guesses into the ciphertext we end up with the partial solution

```
SCEACSKDXACESDSCKVSOLCDDAGKEMGAMTYKTOVKSOSFNCFPCEE
ia  ai    a i ia  itha          e  et  itingan a
XMTDAOLTCQOLGKGACOKSADSFNEGFGNKCHLQHGFOLTMQRITYOSF
  e  thea th    at i  ing  n g a h   nthe    e tin
VLSYLSCFCDXMTGFTLQFPKTPCFPMSWOXMTHCKCOTYSHLTKMRQOS
 hi hiana   e neh n  e an  i t  e a ate i he    ti
YGFATMMOLCOOLSMSMTFOSKTDXFTVOGETOLTGRITYOGAOLGMTVL
  n e  thatthi i enti e  ne t  ethe   e t  th  e h
GSFUTFOTPOLTMXMOTELCMCHHCKTFODXRTTFOGYGFYTCDOLCOOL
 in ente the   te ha a  a ent   eent   n ea thatth
TMTYLCKCYOTKMYGFUTXCETMMCNTCFPOGNSUTOLTSPTCOLCOOLT
e e ha a te    n e a e  agean t gi ethei eathatthe
XCKTOLTETKTKCFPGEMBTOYLTMGAYLSDPKTFCKOLQKYGFCFPGXD
 a ethe e e an     et he    hi   ena th    nan
TOLTCPUTFOQKTGAOLTPCFYSFNETF
ethea  ent  e  the an ing en
```

Of the high frequency letters we still need to assign *o*, *r*, and *s*. We notice that the digraph GF occurs seven times. That represents ?n in plaintext, indicating that the ? is probably a vowel. The only two vowels left are *o* and *u* and the sequence *on* occurs much more frequently in English than *un*, so it is possible that G = *o*. We also see the sequence OLCOOLSMSM, which is currently decrypted as *thatthi?i?* and which might logically decrypt as *that this is* if M = *s*. In addition, there are two double M's in the cryptogram, reinforcing the idea that M = *s*. Finally, for the high-frequency letters we notice that there are 8 KT pairs in the cryptogram. We already know that T = *e* and we also know that *re* is a high-frequency digraph, so its possible that K = *r*. Adding these to the ciphertext we end up with

```
SCEACSKDXACESDSCKVSOLCDDAGKEMGAMTYKTOVKSOSFNCFPCEE
ia  air   a i iar itha   or so se ret ritingan a
XMTDAOLTCQOLGKGACOKSADSFNEGFGNKCHLQHGFOLTMQRITYOSF
 se  thea thoro atri  ing onogra h  onthes   e tin
VLSYLSCFCDXMTGFTLQFPKTPCFPMSWOXMTHCKCOTYSHLTKMRQOS
 hi hiana  seoneh n re an si t se arate i hers  ti
YGFATMMOLCOOLSMSMTFOSKTDXFTVOGETOLTGRITYOGAOLGMTVL
 on essthatthisisentire  ne to etheo  e to those h
GSFUTFOTPOLTMXMOTELCMCHHCKTFODXRTTFOGYGFYTCDOLCOOL
oin ente thes ste hasa  arent   eento on ea thatth
TMTYLCKCYOTKMYGFUTXCETMMCNTCFPOGNSUTOLTSPTCOLCOOLT
ese hara ters on e a essagean togi ethei eathatthe
XCKTOLTETKTKCFPGEMBTOYLTMGAYLSDPKTFCKOLQKYGFCFPGXD
 arethe ereran o s et heso  hi  renarth r onan o
TOLTCPUTFOQKTGAOLTPCFYSFNETF
ethea  ent reo the an ing en
```

This is the breakthrough we needed. The analysis now depends on guessing possible words that we can see hints of in the partially decoded ciphertext. It is easy to see words like *writing*, *message*, *separate*, *secret*, etc. and we can now uncover the plaintext in short order. The final plaintext is (with punctuation added).

> I am fairly familiar with all forms of secret writing, and am myself the author of a trifling monograph upon the subject, in which I analyse one hundred and sixty separate ciphers, but I confess that this is entirely new to me. The object of those who invented the system has apparently been to conceal that these characters convey a message, and to give the idea that they are the mere random sketches of children. Arthur Conan Doyle, "The Adventure of the Dancing Men" [2].

So what is the process of cryptanalysis here? We begin with two facts, the relative frequency counts in English, and the behavior of digraphs and trigraphs as they appear in words in English. Then we get the actual frequency counts in the cryptogram and use our knowledge to try to identify the high-frequency letters and digraphs in the cryptogram. Once we have a partial reconstruction using the high-frequency letters we can then begin to guess whole words, filling in more letter equivalents as we go.

References

1. Caesar, Julius. 2008. *The Gallic Wars*. Hardcover. Oxford, UK: Oxford University Press.
2. Doyle, Sir Arthur Conan. 1903. The adventure of the dancing men. *The Strand Magazine*.
3. Seutonius. 1957. *The Twelve Caesars*. Paperback. Trans Robert Graves. London, UK: Penguin Classics.

Chapter 3
The Black Chambers: 1500–1776

Abstract The period from 1500 through the middle of the 18th century saw the creation of modern nations and city-states. It also saw increased use of codes and ciphers in diplomacy, the military, and commerce. The nomenclator, a marriage of the code and cipher is a product of this period. This period also saw the creation of a cipher that would remain "unbreakable" for 350 years, the polyalphabetic substitution cipher. This chapter traces the history of the Black Chambers, those organizations created by the newly formed nations to break the codes and ciphers of their neighbors, and it describes the nomenclator and the evolution of the polyalphabetic substitution cipher known as the Vigenère cipher.

3.1 Mary, Queen of Scots and the Spymaster

Sir Francis Walsingham had a problem. Her name was Mary Stuart and she was the former Queen of Scotland and heir apparent to the throne of England. She'd been a prisoner of the Queen of England, Elizabeth I, for 18 years and Walsingham, Elizabeth's Principal Secretary and chief spymaster, wanted nothing more than to end Mary Stuart's imprisonment—and not in a good way.

Mary Stuart had become Queen of Scotland in 1542 when she was 6 days old, upon the death of her father, James V. She was a Catholic in an increasingly Protestant country, and after an aborted rebellion in 1548 she was taken to France where she grew up in the royal court. In order to strengthen the ties between France and Scotland and to stymie the English at the same time, Mary was betrothed to the Dauphin Francis, heir to the French throne, when she was six. Growing up together in the French court, Mary and Francis grew to love each other and were married on 24 April 1558 when Mary was nearly sixteen and Francis was fourteen. Shortly thereafter, Francis' father, Henry II of France was killed in an accident in a jousting tournament and Francis became King of France on 10 July 1559, with Mary as his queen consort. In addition to being the King of

J. F. Dooley, *A Brief History of Cryptology and Cryptographic Algorithms*, SpringerBriefs in Computer Science, DOI: 10.1007/978-3-319-01628-3_3,

France, Francis was also the king consort of Scotland because of his marriage to Mary. Unfortunately, Francis II had always suffered from ill health, and shortly after he became king an ear infection that had bothered him since he was a child flared up. An abscess developed on his brain and he died on 5 December 1560 after only seventeen months on the throne. Having been shut out of French politics after Francis' death and with a mother-in-law, Catherine de Medici, who never liked her, Mary returned to Scotland in September 1561.

Mary, who was personable, smart, and somewhat wily in the ways of Scottish politics, was also stubborn, rash, and willful. She ruled Scotland rather peacefully for four years until her marriage to her first cousin, Henry Stuart, the Earl of Darnley. It was only after their marriage that Mary discovered that Darnley was vicious, abusive, ambitious, and cruel. It wasn't long before many of the Scottish nobles, and eventually Mary as well, were plotting ways to "set Darnley aside." It was most likely no surprise when a house where Darnley was staying while he recuperated from an illness blew up the night of 9–10 February 1567. Darnley's body was found, strangled (or smothered—the accounts differ) in the garden. And thus ended Mary's second marriage. The best thing that came out of that was the birth of Mary's only child James on 19 June 1566. It was James who would become James VI of Scotland and, because both his parents were descended from Margaret Tudor, Henry VIII's older sister, also James I of England.

Mary's mistakes in love and politics continued when in May 1567 she married James Hepburn, the Earl of Bothwell, who had just been acquitted of Darnley's murder. This was another ill-considered and ill-fated match as it is believed that Bothwell first abducted Mary, possibly raped her, and then transported her to Edinburgh where they married in a Protestant service. Nobody liked Bothwell. The Protestants in Scotland were shocked that Mary would marry so soon after her husband's death and to the man who was likely involved in Darnley's murder. The Catholics were aghast that Mary would marry in a Protestant service. The whole affair was really the beginning of the end for Mary. By the summer of 1567 the Scottish nobles and Parliament had had enough. Bothwell was exiled to Denmark where he was imprisoned, went insane, and died in 1578. Mary was imprisoned in Loch Leven Castle and on 24 July she was forced to abdicate in favor of her fourteen-month-old son, James. Mary stayed at Loch Leven till the spring of 1568 when, with her jailer's help, she escaped, raised an army of 6,000 and tried to take back her throne. Her royalist forces were soundly defeated on 13 May 1568 at the Battle of Langside, near Glascow. Unable to cross Scotland to take ship for France, Mary fled to England where she asked her cousin, Elizabeth I, for sanctuary and instead ended up in prison.

Eighteen years later, in 1586 Mary was still in prison. Over the years, she had been moved from place to place in England, never close to the sea or to Scotland, and over the years her privileges and freedom had been more and more constrained. She finally ended up at Chartley Hall under the watchful eye of Sir Amias Paulet, a Puritan. She had managed to keep up a correspondence with her agents and sympathizers in France, but by 1584 she was allowed virtually no correspondence. Her letters to her son James were confiscated at the Scottish border and his Protestant uncle, Mary's half-brother the Earl of Moray acting as regent, raised James. James was constantly told that his mother had killed his father and abandoned him, so there was no love lost on his part.

Mary never gave up hope of returning to Scotland and regaining her throne; she also was always aware of her position as heir apparent to the English throne, and this is what finally sealed her fate.

Mary's fortunes seemed to change on 16 January 1586 when she received two letters; one from her agent in Paris, Thomas Morgan, and one from Chateauneuf, the French ambassador to England. A Catholic loyalist, Gilbert Gifford, delivered the letters in a roundabout way. Gifford had been born in England and had studied for the priesthood in Rome and Rheims. He had recently returned to England to help the Catholic cause. He had arranged with a local brewer to hide the letters in a leather pouch, which was inserted into a hollow bung that was then put into a beer barrel. When the barrel was delivered, the bung was removed, the letters extracted, and the bung replaced. Sending letters out of Chartley Hall reversed the process. After the first letters, Mary immediately replied to the French ambassador and enclosed a new cipher for his use because the cipher he had was over two years old. She also warned him about spies—"She begged him, too, to be on strict guard against the spies who, under the color of the Catholic religion, would be assiduously working to penetrate his house, and her secrets, as they had under her predecessor." [1, p. 153]

The latter was good advice that Mary herself should have heeded. It turned out that Gifford was a double agent, working for Sir Francis Walsingham. Gifford had offered his services to Walsingham in the fall of 1585, and had ingratiated himself in the English Catholic clique in England upon his return to England from France in December 1585. After that initial delivery of letters in January, Gifford kept up a regular schedule of visits and carried letters between Mary and the French ambassador and English Catholic conspirators. As he was coming and going, he would make a side-trip and deliver the letters to Thomas Phelippes, Walsingham's cryptographer who would have the letters unsealed, copied, and resealed before their delivery. Mary, having generously and innocently provided the cipher she was using after the first batch of letters, allowed Phelippes to simply decrypt each letter as it arrived, with no cryptanalysis being necessary.

Mary's cipher was a small *nomenclator*, the standard diplomatic and personal cipher system throughout Europe beginning in the Renaissance period. Designed to be more secure than a simple cipher and easier to use than a codebook, they were a combination of a monoalphabetic cipher, sometimes with nulls and homophones, and a small codebook with typically a few hundred codewords, although some were considerably larger. Mary's system was a particularly easy nomenclator to break, having only 23 symbols in the cipher alphabet and 36 codewords in the code part [4, p. 38].

All through the spring and early summer of 1586 Gifford kept up his courier duties while Walsingham and Phelippes watched and waited for a slip that would deliver Mary into their hands. The end game finally began in May when a small group of Catholic royalists began meeting at the Plough Inn near the Temple bar. The head of the conspiracy was Anthony Babington, a twenty-five year old, well-to-do Catholic who had been a page at the Earl of Shrewsbury's house when Mary was a prisoner there. Babington gathered a half a dozen of his friends together and hatched a plot to assassinate Elizabeth and foment a Catholic uprising to put Mary on the throne with the help of troops from Philip II of Spain. Eventually the conspiracy grew to thirteen or more—some of whom were Walsingham's spies.

Meeting through the spring of 1586, the conspirators developed their plans and decided that they couldn't proceed without approval from Mary, Queen of Scots herself. On 7 July Babington wrote a letter to Mary laying out all the details of the conspiracy and gave it to Gilbert Gifford for delivery. The plan was hazy in its details, but was more than enough for Walsingham. According to Budiansky,

> Babington himself would lead ten gentlemen and a hundred followers to 'undertake the delivery of your royal person from the hands of your enemies.' And 'for the dispatch of the usurper, from the obedience of whom we are by excommunication of her made free, there be six noble gentlemen all my private friends who for the seal they bear to the Catholic cause and your Majesty's service will undertake their tragical execution.'[1, p. 160]

Despite this incriminating evidence, Walsingham waited. He wanted Mary's own approval of the plot and proof that she was involved in attempting to assassinate Elizabeth. The confirmation he sought came on 17 July 1586 when Mary replied to Babington, approving the plot, asking for more details, and ending with "The affairs being thus prepared and forces in readiness both within and without the realm, then shall it be time to set the six gentlemen to work, taking order, upon the accomplishing of their design, I may be suddenly transported out of this place…Fail not to burn this quickly." [1, p. 161]. And thus, she sealed her fate. Babington was alarmed and bolted on 4 August. He and most of his conspirators were captured on 15 August, and after a bit of torture and a speedy trial Babington and six of his co-conspirators were hung, drawn, and quartered on 20 September 1586.

Meanwhile, Mary had been arrested on 11 August and on 25 September 46 nobles, including Walsingham, took her to Fotheringhay Castle for a trial. The trial began on 15 October and lasted two days, during which Mary consistently denied all the charges and proclaimed her innocence. But the cipher letters were the most damming evidence presented and even Mary had no answer to them. She was convicted of treason on 25 October and sentenced to death.

At this point Elizabeth began vacillating and looking for a way to approve the execution without it being blamed on her. Finally on 1 February 1587 Elizabeth signed the death warrant. To avoid having Elizabeth change her mind, the order of execution was delivered on 5 February and Mary was beheaded in the Fotheringhay Great Hall on the morning of 8 February 1587. Mary walked regally up the scaffold, forgave her executioners and prayed for her son before the execution. In order to avoid any of Mary's possessions being turned into relics by the English Catholics, all her clothes and even the headsman's block itself were burned.

3.2 Nomenclators

Nomenclators originated in the early Renaissance period as a way to make the monoalphabetic substitution cipher more secure. By the 1400s frequency analysis was a well-known technique of cryptanalyzing monoalphabetic substitutions. It was thought that adding a codebook to the cipher system would make the message harder to cryptanalyze, and this does work, up to a point. Several issues arise with the use of nomenclators. First, the size of the codebook is important. The more codewords involved, the more ciphertext must be intercepted in order to make a break in the code. So over time the codebook part of nomenclators grew. Secondly, because part of the message was still enciphered using a monoalphabetic substitution cipher, the cryptanalyst could still use frequency analysis on that part and attempt to guess the codewords based on context. Thirdly, because a codebook is used, these books must be distributed to all the correspondents, so nomenclators do not eliminate the distribution problem. Finally, with many nomenclators the cipher alphabet doesn't change. So once the substitution cipher part of the nomenclator has been broken, it is broken for good.

Despite these failings, nomenclators became more and more popular in diplomatic and, to a lesser degree, military cryptologic systems from around 1400 up until the early part of the 19th century. As their popularity grew, it became more important to intercept and break them. Just as Walsingham recognized the usefulness of reading an enemies enciphered correspondence, other European city-states and countries did the same. This led, in the late 1500s, to the creation of the *chambres noire*—the Black Chambers housed in the foreign offices of many European countries.

3.3 The Black Chambers

Leading the way were the Italians. With the growth of powerful city-states in Italy, secretaries whose sole occupation was to create and to break cryptograms of other countries and city-states began to appear. By the mid 1600s nearly every nation in Europe had its own Black Chamber, including England, France, Austria-Hungary, the Vatican, Spain, Sweden, Florence, Venice, and Switzerland. In many of these countries the job of cipher secretary was passed on from father to son, giving the names of famous families of cryptographers from the period. Names such as Antoine Rossignol of France, who invented the 2-part nomenclator and whose son and grandson also became cipher secretaries to the French monarch.

In England, the mathematician John Wallis had the distinction of solving cipher messages for both Cromwell's roundheads and for the restored King Charles II; he also helped found the Royal Society of London. Wallis' grandson succeeded him, but met an untimely end only six years into his tenure. Edward Willes replaced him in 1716. Willes proved to be a very competent cryptanalyst and passed the torch on to three of his sons and then to three grandsons. As a result, the Willes clan were the principal cipher secretaries for England through nearly all the 18th century.

The Austrians had the reputation for having the best and most efficient Black Chamber in Europe, and the most democratic. Cryptanalysts worked one week on, one week off and they received bonuses for difficult decipherments. They were recruited from all walks of life with the requirements that they knew some algebra and other mathematics, spoke French and Italian, and were of "high moral caliber." [3, p. 165]

3.4 The Next Complexity: Polyalphabetic Substitution

With the rise and success of the various Black Chambers it became clear that nomenclators were vulnerable to cryptanalysis, making this a period when the cryptanalysts had the upper hand over the cryptographers. So what were cryptographers to do to regain the ascendency and make their secret correspondence secret again? They developed two different methods that enabled the cryptographers to once again have the upper hand; the modern code, and the polyalphabetic substitution cipher.

The monoalphabetic cipher was vulnerable to frequency analysis because it failed to hide the language characteristics of the plaintext language. One way to obscure language features is to remove all word divisions from a cryptogram and just send the ciphertext in equal-sized groups of letters or symbols. This obscures word and sentence features, but does nothing about letter frequencies. The way to obscure letter frequencies is to use more than one cipher alphabet. This then creates more than one substitution letter or symbol for every letter in the plain alphabet. Thus an 'e' could be replaced by an 's' in one place, by a 'k' in another, and by a 'd' in a third, hiding the frequency of occurrence of the 'e'. Such methods flatten the frequency distribution. The more cipher alphabets that are used the more possible substitutions there are for each plaintext letter and the flatter the frequency chart becomes. The flatter chart then makes it harder it to find the cipher letter—plain letter equivalences. All of which makes the cryptanalyst's job even more difficult.

This is the idea that Leon Battista Alberti presented in an essay on cryptography he published in 1466 or 1467. Alberti, born in 1404, was a true Renaissance man who was an architect, poet, musician, philosopher, and a writer of books on architecture, morality, law, painting, and cryptography. In his 1466 essay Alberti described a disk made of two copper plates with each plate divided into 24 sections. On the outer plate 20 letters of the Latin alphabet were inscribed in order. At that time the classical Latin alphabet didn't include the letters J, U, and W and the Italian language did not use H, K, and Y. The final four cells were filled with the numerals 1, 2, 3, and 4. The inner plate used all 23 letters of the classical Latin alphabet and the digraph "et" meaning & in a mixed order. The two plates were laid on top of one another and a spike driven through their centers. Now the inner plate could rotate. Alberti used the outer plate as the plain alphabet and the inner as the cipher alphabet. His enciphering procedure was to choose a single index letter on the inner plate and rotate it till it appeared under some random letter on the outer plate. This then gave Alberti a single mixed cipher alphabet. The encipherer would then write the random letter down on the message and then proceed to encipher several words using the same alphabet. He would then move the index letter until it was under some other letter (a new random letter) on the outer plate and proceed to encipher several more words with this new mixed cipher alphabet. This continued until the entire message was enciphered. Alberti's method was ingenious and was the first time that a description of a system that used more than one cipher alphabet was used. But it didn't use a key word, and it enciphered large groups of consecutive letters using the same alphabet.

The next improvement in the polyalphabetic cipher came about fifty years later in 1518 with the posthumous publication of Johannes Trithemius' book *Polygraphie*. Trithemius' contribution was to publish the first polyalphabetic square or tableau. Trithemius' *tabula recta* was the simplest of all, just using the 26 alphabets of the Caesar standard shift as shown in Table 3.1.

Trithemius enciphered a text by using the cipher alphabet in the first row for the first letter, the cipher alphabet in the second row for the second letter, etc. all the way to the bottom and then beginning again with the top row. He did not use a key or a keyword. Giovan Batista Belaso would introduce that next improvement in 1553.

With the idea of a keyword, all the parts of a modern polyalphabetic system were in place. It took another Italian, Giovanni Batista Porta to put all the ideas together. In his essay *De Furtivis Literarum* in 1563, Porta used the idea of a mixed alphabet from Alberti, Trithemius' square and letter-by-letter alphabet change, and Belaso's keyword to create a single system for polyalphabetic substitution. Alas, with the vagaries of history Porta is not usually credited with this clever synthesis of ideas. That credit goes to someone who had nothing to do with the creation of the polyalphabetic substitution system, but who actually invented a more secure version of the system—for which he gets no credit.

Blaise de Vigenère was born on 5 April 1523. At the age of twenty-two he entered the diplomatic service and it was during a two-year posting to Rome in 1549 that he became immersed in cryptology. Retiring from diplomatic service in 1570 at the age of 47, he devoted the rest of his life to writing. His most famous book, and the one that ensures his place in cryptologic history, is his 1585 *Traicté des Chiffres*. The most important part of this book—and the part for which he gets no credit—is his development of the *autokey* cipher. In Vigenère's autokey, there is a priming key, a single letter that is used as the key to encrypt the first letter of the plaintext. The rest of the key is the plaintext itself, so the second letter of plaintext uses the first letter of plaintext as it's key letter. Similarly, the third letter of plaintext uses the second plaintext letter as it's key letter, etc. This system is much more secure than any of Alberti's, Trithemius' or Porta's systems. Interestingly, the autokey system was forgotten for nearly 300 years, only to be resurrected in the late 19th century. What Vigenère *does* get credit for is the polyalphabetic system that uses standard alphabets and encrypts letter by letter using a short, repeating keyword; one of the simplest polyalphabetics to solve.

Table 3.2 shows what is now known as the Vigenère tableau.

Table 3.1 Johannes Trithemius' *tabula recta*

A	B	C	D	E	F	G	H	I	J	K	L	M	N	O	P	Q	R	S	T	U	V	W	X	Y	Z
B	C	D	E	F	G	H	I	J	K	L	M	N	O	P	Q	R	S	T	U	V	W	X	Y	Z	A
C	D	E	F	G	H	I	J	K	L	M	N	O	P	Q	R	S	T	U	V	W	X	Y	Z	A	B
D	E	F	G	H	I	J	K	L	M	N	O	P	Q	R	S	T	U	V	W	X	Y	Z	A	B	C
E	F	G	H	I	J	K	L	M	N	O	P	Q	R	S	T	U	V	W	X	Y	Z	A	B	C	D
F	G	H	I	J	K	L	M	N	O	P	Q	R	S	T	U	V	W	X	Y	Z	A	B	C	D	E
G	H	I	J	K	L	M	N	O	P	Q	R	S	T	U	V	W	X	Y	Z	A	B	C	D	E	F
H	I	J	K	L	M	N	O	P	Q	R	S	T	U	V	W	X	Y	Z	A	B	C	D	E	F	G
I	J	K	L	M	N	O	P	Q	R	S	T	U	V	W	X	Y	Z	A	B	C	D	E	F	G	H
J	K	L	M	N	O	P	Q	R	S	T	U	V	W	X	Y	Z	A	B	C	D	E	F	G	H	I
K	L	M	N	O	P	Q	R	S	T	U	V	W	X	Y	Z	A	B	C	D	E	F	G	H	I	J
L	M	N	O	P	Q	R	S	T	U	V	W	X	Y	Z	A	B	C	D	E	F	G	H	I	J	K
M	N	O	P	Q	R	S	T	U	V	W	X	Y	Z	A	B	C	D	E	F	G	H	I	J	K	L
N	O	P	Q	R	S	T	U	V	W	X	Y	Z	A	B	C	D	E	F	G	H	I	J	K	L	M
O	P	Q	R	S	T	U	V	W	X	Y	Z	A	B	C	D	E	F	G	H	I	J	K	L	M	N
P	Q	R	S	T	U	V	W	X	Y	Z	A	B	C	D	E	F	G	H	I	J	K	L	M	N	O
Q	R	S	T	U	V	W	X	Y	Z	A	B	C	D	E	F	G	H	I	J	K	L	M	N	O	P
R	S	T	U	V	W	X	Y	Z	A	B	C	D	E	F	G	H	I	J	K	L	M	N	O	P	Q
S	T	U	V	W	X	Y	Z	A	B	C	D	E	F	G	H	I	J	K	L	M	N	O	P	Q	R
T	U	V	W	X	Y	Z	A	B	C	D	E	F	G	H	I	J	K	L	M	N	O	P	Q	R	S
U	V	W	X	Y	Z	A	B	C	D	E	F	G	H	I	J	K	L	M	N	O	P	Q	R	S	T
V	W	X	Y	Z	A	B	C	D	E	F	G	H	I	J	K	L	M	N	O	P	Q	R	S	T	U
W	X	Y	Z	A	B	C	D	E	F	G	H	I	J	K	L	M	N	O	P	Q	R	S	T	U	V
X	Y	Z	A	B	C	D	E	F	G	H	I	J	K	L	M	N	O	P	Q	R	S	T	U	V	W
Y	Z	A	B	C	D	E	F	G	H	I	J	K	L	M	N	O	P	Q	R	S	T	U	V	W	X
Z	A	B	C	D	E	F	G	H	I	J	K	L	M	N	O	P	Q	R	S	T	U	V	W	X	Y

The top row of the table is the plaintext alphabet and the leftmost column is the key alphabet. In this system, of course, both correspondents must know the keyword. The encipherer takes the next letter from the keyword to select the row to use. The plaintext letter is selected from the appropriate column of the top row and the intersection of the row and the column is the ciphertext letter. If the key is TURING and the plaintext is "Alan was not the only person to be thinking about mechanical computation…" then for the first few letters we would get

```
Key:    T U R I N G T U R I N G T U R I N G
Plain:  a l a n w a s n o t t h e o n l y p
Cipher: T F R V J G L H F B G N X I E T L V
```

Because we are using standard shifted alphabets we can simplify the work by using a little modular arithmetic. If we were to number the letters of the alphabet so that A = 0, B = 1, C = 2, etc. down to Z = 25 then encryption using a Vigenère cipher could be expressed mathematically as

$$C_i = (P_i + K_j) \bmod 26$$

Table 3.2 A modern Vigenère tableau

	a	b	c	d	e	f	g	h	i	j	k	l	m	n	o	p	q	r	s	t	u	v	w	x	y	z
A	A	B	C	D	E	F	G	H	I	J	K	L	M	N	O	P	Q	R	S	T	U	V	W	X	Y	Z
B	B	C	D	E	F	G	H	I	J	K	L	M	N	O	P	Q	R	S	T	U	V	W	X	Y	Z	A
C	C	D	E	F	G	H	I	J	K	L	M	N	O	P	Q	R	S	T	U	V	W	X	Y	Z	A	B
D	D	E	F	G	H	I	J	K	L	M	N	O	P	Q	R	S	T	U	V	W	X	Y	Z	A	B	C
E	E	F	G	H	I	J	K	L	M	N	O	P	Q	R	S	T	U	V	W	X	Y	Z	A	B	C	D
F	F	G	H	I	J	K	L	M	N	O	P	Q	R	S	T	U	V	W	X	Y	Z	A	B	C	D	E
G	G	H	I	J	K	L	M	N	O	P	Q	R	S	T	U	V	W	X	Y	Z	A	B	C	D	E	F
H	H	I	J	K	L	M	N	O	P	Q	R	S	T	U	V	W	X	Y	Z	A	B	C	D	E	F	G
I	I	J	K	L	M	N	O	P	Q	R	S	T	U	V	W	X	Y	Z	A	B	C	D	E	F	G	H
J	J	K	L	M	N	O	P	Q	R	S	T	U	V	W	X	Y	Z	A	B	C	D	E	F	G	H	I
K	K	L	M	N	O	P	Q	R	S	T	U	V	W	X	Y	Z	A	B	C	D	E	F	G	H	I	J
L	L	M	N	O	P	Q	R	S	T	U	V	W	X	Y	Z	A	B	C	D	E	F	G	H	I	J	K
M	M	N	O	P	Q	R	S	T	U	V	W	X	Y	Z	A	B	C	D	E	F	G	H	I	J	K	L
N	N	O	P	Q	R	S	T	U	V	W	X	Y	Z	A	B	C	D	E	F	G	H	I	J	K	L	M
O	O	P	Q	R	S	T	U	V	W	X	Y	Z	A	B	C	D	E	F	G	H	I	J	K	L	M	N
P	P	Q	R	S	T	U	V	W	X	Y	Z	A	B	C	D	E	F	G	H	I	J	K	L	M	N	O
Q	Q	R	S	T	U	V	W	X	Y	Z	A	B	C	D	E	F	G	H	I	J	K	L	M	N	O	P
R	R	S	T	U	V	W	X	Y	Z	A	B	C	D	E	F	G	H	I	J	K	L	M	N	O	P	Q
S	S	T	U	V	W	X	Y	Z	A	B	C	D	E	F	G	H	I	J	K	L	M	N	O	P	Q	R
T	T	U	V	W	X	Y	Z	A	B	C	D	E	F	G	H	I	J	K	L	M	N	O	P	Q	R	S
U	U	V	W	X	Y	Z	A	B	C	D	E	F	G	H	I	J	K	L	M	N	O	P	Q	R	S	T
V	V	W	X	Y	Z	A	B	C	D	E	F	G	H	I	J	K	L	M	N	O	P	Q	R	S	T	U
W	W	X	Y	Z	A	B	C	D	E	F	G	H	I	J	K	L	M	N	O	P	Q	R	S	T	U	V
X	X	Y	Z	A	B	C	D	E	F	G	H	I	J	K	L	M	N	O	P	Q	R	S	T	U	V	W
Y	Y	Z	A	B	C	D	E	F	G	H	I	J	K	L	M	N	O	P	Q	R	S	T	U	V	W	X
Z	Z	A	B	C	D	E	F	G	H	I	J	K	L	M	N	O	P	Q	R	S	T	U	V	W	X	Y

where C_i is the ith ciphertext letter, P_i is the ith plaintext letter, and K_j is the jth key letter. We have to use a different index for the key because it is short and repeats throughout the plaintext encipherment. So in the example above, we would have

19 = (0 + 19) mod 26 (a maps to T using key letter T),
05 = (11 + 20) mod 26 (l maps to F using key letter U),
17 = (0 + 17) mod 26 (a maps to R using key letter R), etc.

With the advent of the complete polyalphabetic substitution cipher system the cryptographers had the upper hand once again. By using multiple alphabets the system flattened out the frequency chart, eliminating the best opportunity the cryptanalyst had for solving the cryptogram.

For example, if we use the following text

> Alan was not the only person to be thinking about mechanical computation in nineteen thirty-nine. There were a number of ideas and initiatives, reflecting the growth of new electrical industries. Several projects were on in the United States…In the normal course of events Alan could have expected fairly soon to be appointed to a university lectureship,

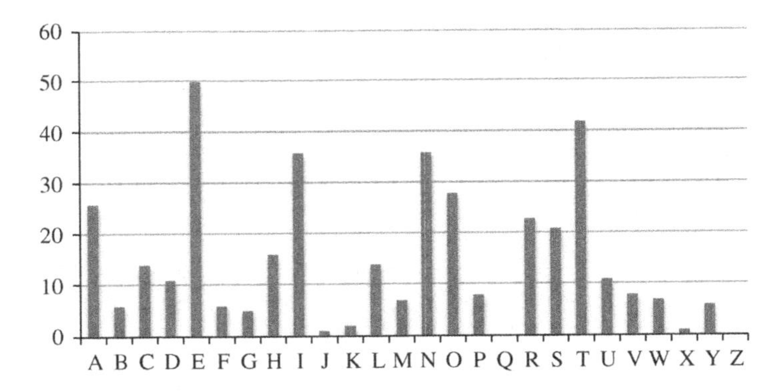

Fig. 3.1 Plaintext frequency chart for Turing quote

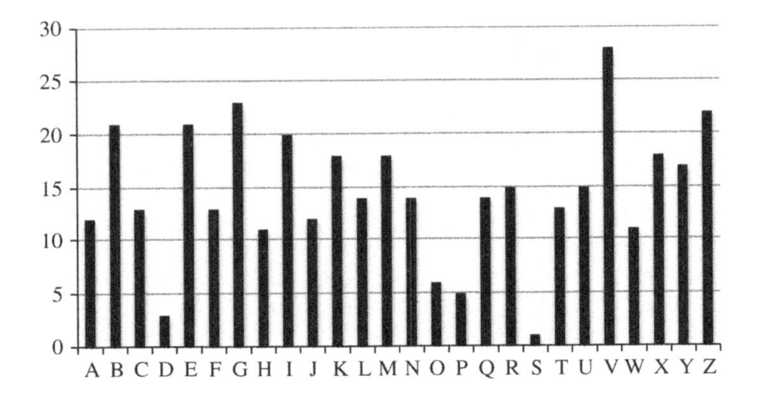

Fig. 3.2 Frequency chart for Turing quote ciphertext

> and most likely to stay on at Cambridge forever. But this was not the direction in which his spirit moved. [2, pp. 155, 157]

We would have a frequency chart that looks like Fig. 3.1.

Now, if we encrypt it using a Vigenère cipher and the keyword TURING we have a frequency chart of the ciphertext that looks like Fig. 3.2.

Notice how the counts have evened out. The distinctive 'E' is not there, nor is the distinctive triple of 'RST', or the dips for 'Z', 'J', and 'Q'. These characteristics are what spelled the eventual doom of the nomenclator because they made the Vigenère cipher more secure than the usual nomenclator. Why, then, did the nomenclator continue to be used for another 200 years? It was because the Vigenère was more complicated to use and thus more error-prone. Time and again, organizations would abandon use of the Vigenère because it took too long to create cipher messages and errors in encipherment or decipherment made the ciphertext unreadable [3, p. 150]. But governments continued to try to use it because it was for more than 200 years *le chiffre indéchiffrable*—the undecipherable cipher.

References

1. Budiansky, Stephen. 2005. *Her Majesty's Spymaster*. New York, NY: Penguin Group (USA).
2. Hodges, Andrew. 1983. *Alan turing: The Enigma*. New York: Simon and Schuster.
3. Kahn, David. 1967. *The codebreakers; The story of secret writing*. New York: Macmillan.
4. Singh, Simon. 1999. *The code book: The evolution of secrecy from Mary, Queen of Scots to quantum cryptography*. New York: Doubleday.

Chapter 4
Crypto Goes to War: 1861–1865

Abstract The 19th century marked the beginning of the use of technology in many areas, and cryptology was no exception. The invention of the telegraph and its rapid and easy communication ushered in the twilight of traditional forms of cryptography. It also marked the beginning of a century and a half of rapid development of new techniques in both cryptography and cryptanalysis, all starting during the American Civil War. This chapter looks at the cipher systems used by both the Union and Confederate sides during the American Civil War. It also presents a description of the biggest cryptanalytic breakthrough of the 19th century, the breaking of the unbreakable cipher, the Vigenère.

4.1 Technology Goes to War

By 1861, despite having only been available for about 25 years, the telegraph was nearly ubiquitous in the United States. Its ease and rapidity of communication made it the logical choice for military communications and it changed the face of communications in the military; in short order the telegraph caused both the Union and Confederate forces in the American Civil War (1861–1865) to rethink their use of traditional codes. There were at least two good reasons to make the switch. First, codes were hard to use in the field. Codebooks could be easily lost and would then have to be re-issued to every command. Second, the advent of the telegraph had turned command posts into telegraph communication centers and increased the volume of traffic enormously. Because it was easy to string telegraph lines commanders were able to issue increasingly detailed and tactical orders to lower level forces. This increased the number of codebooks that must be printed and distributed; and if a book was captured, it increased the time and effort involved in changing codes. Ciphers were much easier from a tactical viewpoint. Thus, *field ciphers* were born [7, p. 191].

J. F. Dooley, *A Brief History of Cryptology and Cryptographic Algorithms*, SpringerBriefs in Computer Science, DOI: 10.1007/978-3-319-01628-3_4,

4.2 The Union Tries a Route

During the American Civil War, General Edward Porter Alexander, a commander of artillery, was the father and commander of the Confederate Army Signal Corps. It was Alexander who set up the Confederate States telegraph operations, helped design their cryptographic systems, and tried to decrypt Union correspondence. He was also the artillery officer in charge of the bombardment before Pickett's Charge on the last day of the Battle of Gettysburg. One night in 1863, Alexander was handed a Union cryptogram that had been taken from a courier who had been captured near Knoxville, Tennessee. The cryptogram read

> To Jaque Knoxville, Enemy the increasing they go period this as fortified into some be it and Kingston direction you up cross numbers Wiley boy Burton and if will too in far strongly go ought surely free without your which it ought and between or are greatly for pontoons front you we move as he stores you not to delay spare should least to probably us our preparing Stanton from you combinedly between to oppose fortune Roanoke rapid we let possible speed if him that and your time a communication can me at this news in so complete with the crossing keep move hear once more no from us open and McDowell Julia five thousand ferry (114) the you must driven at them prisoners artillery men pieces wounded to Godwin relay horses in Lambs (131) of and yours truly quick killed Loss the over minds ten snow two deserters Bennet Gordon answer also with across day (152).

According to Alexander, "I had never seen a cipher of this character before, but it was very clear that it was simply a disarrangement of words, what may be called, for short, a jumble" [6, p. 111].

And a jumble it was. After spending the entire night trying to unscramble the jumble, Alexander gave up; he was never able to decipher the Union message. What Alexander had come up against was the Union Army's main command cipher, used between generals and between the Union Armies and Washington. A telegrapher who had started the war working for the Governor of Ohio designed it. It was during that time he produced a simple cipher for the Governor's use that allowed him to send secret correspondence to the Governors of Indiana and Illinois. That telegrapher, who would help found the Western Union Company and be the first president of the Western Electric Manufacturing Company, was Anson Stager.

The cipher that Stager created in 1861 started out as a simple route word transposition cipher. In a route word transposition cipher, the plaintext is written out by words in a rectangle, line by line. The plaintext is then taken off by columns, but there is a key that tells the encipherer three things: first, the size of the rectangle to use, second, the order in which to take off each column, and third, the direction—up or down—in which to take off the words. For example, if the message is

Table 4.1 Sample message rectangle

the	enemy	has	changed
his	position	during	the
night	deserters	say	that
he	is	retreating	smith

The enemy has changed his position during the night. Deserters say that he is retreating. Smith.

And the rectangle is a 4 × 4, then the plaintext is written out as in Table 4.1.

Then if the code words are taken off in the following order first column down, fourth column up, second column down, and third column up, the resulting cryptogram is

the his night he Smith that the changed enemy position deserters is retreating say during has

this is not the most secure cipher ever invented, but Stager added a few twists that helped make it stronger. First, he added nulls at regular intervals to confuse the Confederate cryptanalysts. So if the words *attacking*, *summer*, *unchanged*, and *him* are nulls (called *blind words* during the Civil War) and are added every four words, the cryptogram changes to

the his night he attacking Smith that the changed summer enemy position deserters is unchanged retreating say during has him

which spreads the words of the ciphertext out a bit and also provides a check for the decipherer that the ciphertext is correct. This last point was important because most of these messages were sent by telegraph and preventing garbled messages was essential. Stager next added a small set of codewords to further hide the identity of people and places and certain actions from the cryptanalyst. Finally, every route transposition cryptogram began with a *commencement word* that told the telegraph operator who would decipher the message the size of the rectangle and the route for the columns [1].

In the beginning of the war, all these rules for Cipher No. 1 fit on a 3 × 5 file card. By the end of the war when Cipher No. 4 was released (the ciphers were released out of numerical order) the description was printed in a 48-page booklet and had 1,608 codewords in it. Table 4.2 shows an example of the list of *commencement* and codewords (at the time called *arbitraries*) for Cipher No. 1.

The first column of the table lists the *commencement words*, with the number being the number of lines in the message—the number of rows in the rectangle. The second column contains the *nulls* or *blind words*. The next two pairs of columns are the coded words and their meanings. For example, *Egypt* is the codeword for General George McClellan. A sample telegram using this system [2] looks like

Table 4.2 The codewords, nulls, and indicators for a Stager cipher

Commencement words		Arbitrary words			
Cipher words					
1 Mail	Check	Scott	Bagdad	Dennison	London
2 May	Charge	McClellan	Mecca	Curtin	Vienna
3 August	Change	Steedman	Bremen	Private	Star
4 March	Cheap	Kelly	Berlin	Bird's Pt	Uncle
5 June	Church	Yates	Dublin	Columbus, Ky	Danube
6 April	Caps	Battes	Turin	Memphis	Darien
7 July	Show	Morris	Venice	Paducah	Darby
8 Telegraph	Sharp	Cox	Brussels	Mound City	Geneva
9 Marine	Shave	Washington	Nimrod	Navy Yard	Mexico
10 Board	Shut	Parkersburg	Cain	Pillow	Brazil
11 Account	Ship	Cornwallis	Abel	Ben. M'Cullough	Grenada
12 Director	Shields	Smithton	Kane	Fremont	Paris
13 President	Poles	Clarksburg	Noah	Hunter	Moscow
14 Central	Tools	Grafton	Lot	Grant	Arabia
15 January	Glass	Cumberland	Jonah	Gen. Smith	Baltic
16 Buffalo	Pet	Wheeling	Peter	Gen. Payne	Britain
17 Pittsburg	Vile	Fairmount	Paul	Gen. McClellan	Egypt
18 Cleveland	Base	Horner's Ferry	Judas	Gen. Allen	Negro
19 Rochester	Miscreant	Cumberland	Job		
20 Audit	Scoundrel	Martinsburg	Joe		
21 Company	Scamp	Richmond	Frank		
22 Station	Thief	Cairo	Sam		
23 Report	Puppy	St. Louis	Ham		
24 December	Gentleman	Marietta	Shem		
25 Boston	Nobleman	Prentiss	Mary		
26 Balance	Just	Lyon	France		
27 Refund		Blair	Rome		
28 Debtor		Pope	Naigara		
29 Creditor		Morton	Peru		
30 Abstract					
31 United					
32 Annual					
33 Duplicate					
No. Lines					

Cain, Va., June 1, 1861
To Egypt, Cincinnati, Ohio:
Telegraph the have be not I hands profane right hired held must start my cowardly to an responsible Crittenden to at polite ascertain engine for Colonel desiring demands curse the to success by not reputation nasty state go of superseded Crittenden past kind of up this being Colonel my just the road division since advance sir kill.
(Signed) F. W. Lander.

Table 4.3 Route transposition rectangle

1	2	3	4	5	6	7
the	have	be	not	I	hands	profane
right	hired	held	must	start	my	cowardly
to	an	responsible	Crittenden	to	at	polite
ascertain	engine	for	Colonel	desiring	demands	curse
the	to	success	by	not	reputation	nasty
state	go	of	superseded	Crittenden	past	kind
of	up	this	being	Colonel	my	just
the	road	division	since	advance	sir	kill

The receiving telegraph operator would begin the decryption by noting that the commencement word is Telegraph, indicating 8 lines in the rectangle. He also knows that the nulls occur every seventh word. For this cryptogram with 56 total words (less the Telegraph) he will therefore create a rectangle with eight rows and seven columns (the last column for the nulls) as in Table 4.3.

The message will then be read off by columns in the order (also specified by the commencement word Telegraph) up the sixth column, down the first, up the fifth, down the second, up the fourth, and down the third. This produces the following plaintext message:

Petersburg, Va. June 1, 1861
To: General G. McClellan
Sir: My past reputation demands at my hands the right to ascertain the state of the advance. Colonel Crittenden not desiring to start, I have hired an engine to go up road. Since being superseded by Colonel Crittenden, must not be held responsible for success of this division.

During the course of the Civil War the U.S. Military Telegraph Department (USMT) that Stager headed released ten different Stager ciphers. As far as is known, the Confederates never broke any of them.

4.3 Crypto for the Confederates

While the Union forces used a simple, but relatively secure cipher system, the Confederate States of America chose what should have been the most secure system at that time for their secret correspondence, the Vigenère cipher system.

In the 300 or so years since Porta had first described the polyalphabetic substitution system, no one had been able to break the system reliably. There were occasional breaks, mostly either through luck, context, or betrayal of the key, but there was no systematic cryptanalytic attack that had been developed. So, while the Vigenère was somewhat difficult to use and prone to errors, particularly when sent over the telegraph, it should have been a very secure system for the Confederates. But the Union cryptanalysts could regularly break messages in the Confederate Vigenère cipher system. Why?

The Confederate cipher system was insecure not because of the system itself, but because of *how it was used* by the Confederate Army. There are three reasons why the Confederates themselves made the system less secure. First, they kept word divisions in the cryptograms. This basic enciphering mistake made it much easier for the Union cryptanalysts to guess probable words in the ciphers. It also allowed them to guess parts of the key word or phrase more easily.

Second, the Confederates only enciphered part of each message, leaving the rest of the message in the clear. While this may appear to make the cipher stronger because there is less ciphertext for the cryptanalyst to work with in each message, this decision gave the Union cryptanalysts the context in which the ciphertext was created, once again allowing them to more easily guess probable words and parts of the key. Finally, it appears that throughout the war that the Confederates used only three keys for the command level version of the cipher, and one of those keys was only introduced in the waning days of the conflict. The keys were COMPLETE VICTORY, MANCHESTER BLUFF, and late in the war, COME RETRIBUTION. Note that all three keys are fifteen letters long, making it even easier for the Union cryptanalysts to produce solutions. Other keys were used at the department level (for the army's purposes, a department was generally a geographic region). For example, in 2006 Boklan deciphered a lost Confederate telegram that used the key BALTIMORE [4].

Given that the Confederates made the job of the Union cryptanalysts easier and basically ruined the security of the Vigenère cipher, it still doesn't answer the fundamental question. How does one solve a polyalphabetic cipher?

4.4 Solving a Vigenère Cipher

When you use a Vigenère cipher to encrypt a message, you use the standard Vigenère table with its 26 shifted standard alphabets, and a key word or phrase that repeats for the entire length of the message. Note that you can also use a set of mixed alphabets with a Vigenère and it only makes the solution a little harder to accomplish. Using a keyword or phrase causes you to use a different alphabet for every letter that is enciphered. This is both the strength and the weakness of the Vigenère system.

In the middle of the 19th century, two different men in two different countries both hit upon the basic flaw in this system that allowed them to create an attack that could reliably break a Vigenère cipher.

Charles Babbage was a well-to-do member of British society. He was intelligent, well read and well educated. He was the eleventh Lucasian professor of mathematics at the University of Cambridge, and he had a number of brilliant and interesting ideas. His only problem was follow-through. Babbage hardly ever finished anything, particularly his Difference and Analytical Engines. Babbage worked on these two devices for decades, and the brilliant ideas behind them are echoed in modern computers. Unfortunately, he never finished either. This is not to say he didn't accomplish many things. He invented the cowcatcher for railroad trains and the speedometer. He contributed to several areas of mathematics including algebra, the calculus of functions, geometry, operations research, and infinite series. And in 1854, to satisfy a bet, Charles Babbage developed a technique for breaking polyalphabetic cipher systems.

The second gentleman who independently discovered how to break polyalphabetics was in many ways the polar opposite of Charles Babbage. Major Friedrich Wilhelm Kasiski enlisted in the Prussian army in 1822 at the age of 17 and spent his entire career in the army. He retired in 1852 and except for a short stint in the 1860s as the commander of the Prussian equivalent of a National Guard battalion he spent most of his retirement writing [7, p. 207]. His most famous book was *Die Geheimschriften und die Dechiffrir-kunst* ("Secret Writing and the Art of Deciphering"), published in 1863. Most of this book is taken up with Kasiski's description of how to break a Vigenère cipher.

What both Babbage and Kasiski realized is that the repetition of the key in a Vigenère ciphertext is the weak link in the cipher. Their brilliant idea was that, given a sufficiently long ciphertext it was possible that identical parts of the plaintext would have been enciphered with the same part of the key, yielding identical ciphertext at two or more places in the enciphered message. They also realized that if one counted the letters from the beginning of the first identical plaintext section to the beginning of the second, that the resulting count would be a multiple of the key length. For a contrived example, if the plaintext is "the codes in the word and the message", and the key is "crypt", then we'd get the following ciphertext

```
Plain:  t h e c o d e s i n t h e w o r d a n d t h e m
Key:    c r y p t c r y p t c r y p t c r y p t c r y p
Cipher: V Y C R H F V Q X G V Y C L H T U Y C W V Y C B
```

Note that the ciphertext pattern VYC occurs three times in the ciphertext, and each time there is a distance of 10 letters between the beginning of one VYC and the next. This happens because the same pattern of plaintext, "the", lines up with the same part of the key, "cry", each time, resulting in the same ciphertext. The repetition of the keyword is the liability here. The ciphertext duplicates are all 10 letters apart. Babbage and Kasiski reasoned that this implies the length of the key is a factor of 10. The factors of 10 are 1, 2, 5, and 10. One could argue that a key of length 2 is too short to provide much security, so that a key of length 5 is more reasonable.

A key of length 5 means that the 1st, 6th, 11th, 16th, etc. letters are all enciphered with the same key letter and hence the same alphabet from the Vigenère table. Similarly, the 2nd, 7th, 12th, 17th, etc. letters are all enciphered with the next key alphabet. So if we break up the cryptogram into 5 groups of letters we then have 5 monoalphabetic substitution ciphertexts. We can then do a frequency analysis of each group and solve each group separately. And in a standard shifted alphabet as in the normal Vigenère table, if we can find a single cipher alphabet letter we then have the entire alphabet. This method is pretty universally known as the Kasiski method.

Why, if Charles Babbage discovered the same method as Major Kasiski and discovered it 9 years earlier, isn't Babbage's name on the method instead? There are two theories for this. First, Babbage was doing this to satisfy a disagreement with a friend, so he didn't really see the impact of a general method for solving the polyalphabetic substitution and he just never considered publishing his results. Given Babbage's history of not following through on some of his work, this is plausible. The second possibility is that Babbage was working on the solution of the Vigenère cipher at the beginning of the Crimean War (1853–1856) and the British government asked him to refrain from publishing the method so that they could use it against secret Russian communications. Although, it is not clear how much the Russians used the Vigenère cipher [8, p. 78]. Regardless, Babbage did not publish and Kasiski did, so it is now the *Kasiski method*.

Finding the key length and then the key is clearly a useful method for solving a polyalphabetic cipher. But there are a couple of problems, as well. Either there may be no repeated ciphertext sections because either the ciphertext is too short, or the key is too long. So having a technique that gave the cryptanalyst the key length without having to search for duplicate sections of ciphertext would be more efficient. That is exactly what William F. Friedman developed in 1920. Friedman, who we will cover in more depth later, was the head of the Cipher Department at the Riverbank Laboratories in Illinois at the time and had already developed several other solutions for various cryptographic problems. His technique for finding the key length in a polyalphabetic substitution cipher, though, was a brilliant breakthrough and was the event that set the science of cryptology on firm statistical ground for the first time [3, p. 76–84].

The following derivation follows Bauer [3, p. 76–78]. Friedman's observation was that, first, you could compute the probability that two randomly chosen letters in a cryptogram would be the same by using the frequency count of that letter in the cryptogram. So if a cryptogram has N letters in it, and say, the As have a frequency of F_A, then the probability that you'd randomly pick an A is $P(A) = F_A/N$. If you then pick a second letter randomly the probability that it will be an A is $P(A_2) = (F_A-1)/(N-1)$. And the probability that you'll pick two random letters that are both A's is just the product of the two or

$$P(A) * P(A_2) = F_A/N * (F_A - 1/(N - 1)$$

Since you could have picked any letter, say D or Q, instead of A, you can create the probability that any two randomly selected letters are the same by summing up the probabilities for each letter. This leads to Friedman's famous definition

$$Index\ of\ Coincidence = \frac{\sum_{i=A}^{Z} F_i(F_i - 1)}{N(N - 1)}$$

This value has a number of characteristics. For a single alphabetic substitution, the value is about 0.066, and for many alphabets—effectively just a random replacement of letters, the value is about 0.038. The value also will change somewhat with

Table 4.4 Expected values for the index of coincidence

Alphabets	Index of coincidence
1	0.0660
2	0.0520
3	0.0473
4	0.0450
5	0.0436
6	0.0426
7	0.0420
8	0.0415
9	0.0411
10	0.0408

the length of the cryptogram. And it will, of course, change based on the contents of each cryptogram, and the value will also vary because of the letter frequency of the language used in the cryptogram. So looking at the expected values for a small number of alphabets and cryptograms in English we get a table that looks like Table 4.4.

So what Friedman had devised was a way to statistically "guess" at the length of the key in a polyalphabetic substitution without having to count the duplicated ciphertext as in the Kasiski method.

This is not to say that we can now abandon the Kasiski method. It turns out that hardly ever do you get the expected value when you compute the index of coincidence, so at best, you have to test two key lengths to see if you've found the correct one. Also, if the key contains two letters that are the same, like CRYPTOLOGY, then that can throw off the count. So the best method is to use *both* the index of coincidence and the Kasiski method, using one method to confirm your guess as to the key length made by the other. The fundamental beauty of the index of coincidence is that it marks the point in time where cryptanalysis (a term that Friedman coined) is firmly grounded in mathematics. From this point on, the new methods developed are fundamentally mathematical, rather than linguistic.

Now lets do an example that demonstrates the use of both the index of coincidence and the Kasiski method. We're given the following cryptogram:

```
KKGHM VGJRG TBIVQ IVWRY CGBSX VPTGQ QLLIX FGUQP BROII
TXBVY CHMFC EETLH KVTTK VGRPS HTKYY KXGGV LWNBF ICSLC
HTGEA STJFJ GRTVB HLSEI CIVWR YCGLC HKFTL HCKCS XDCIR
BXBVJ CRKSV DGABH CIWRB DJVPH MVGVJ PUCTR RTHFK VLITZ
EZNWX FUJWB UCNJS HXRKE ADFAG IAXTZ VIYCL OEKGD GGEZN
WXFUL QTWPA TPXFW PRJHT BFVTT KMUGC RBSUF DBTZG WYRMC
TRLSX JGIWD GSQWR WXAER LQXGQ CTTWK KKFIB AGRLS IOVZC
CVSCE BPEWV KJTDB QNJTW UGFDI ASULZ YXQVS SIMVK JMCXV
GJYIE CQBGC ZOVZR LBHJV WTLVC CDREC UVBIA WUFLT BGVFM
HBARC C
```

This cryptogram contains 411 letters. It's frequency chart looks like Fig. 4.1.

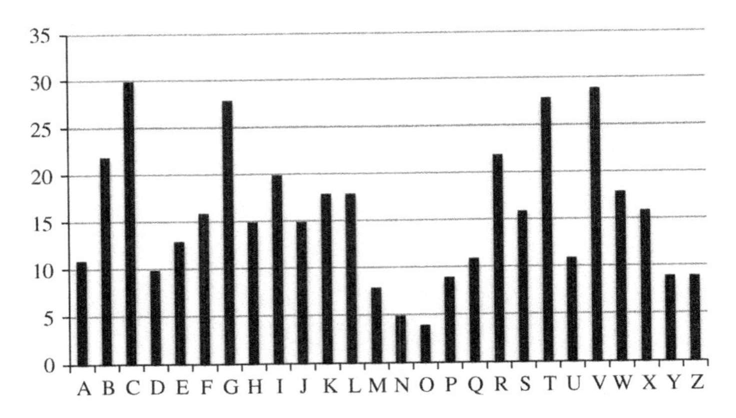

Fig. 4.1 Frequency chart of example Vigenère cryptogram

This frequency table, while not completely flat, is certainly not the frequency chart of a monoalphabetic substitution cipher. Going one step further, the index of coincidence of this ciphertext is 0.0441. Looking at the table of expected values we see that this ciphertext should have been encrypted with a key of length 4 or 5. If we break up the ciphertext into groups of 4 by choosing the 1st, 5th, 9th, etc. and then the 2nd, 6th, 10th, etc. letters and if we compute the index of coincidence on each of those groups, then we should get a set of numbers near 0.066 if our key length guess is correct. Doing this we get for a key length of 4, the values 0.0531, 0.0514, 0.0461, and 0.0460, none of which are very close to 0.066. If we try a key length of 5, we get 0.0444, 0.0458, 0.0397, 0.0473, and 0.0419, again not close to 0.066. Is the index of coincidence just wrong? Lets try a different tack and attempt the Kasiski method. Using the Kasiski method, we need to select a string length, say 4, and beginning at the start of the cryptogram, look for sequences of ciphertext of length 4 that repeat further down the cryptogram. According to Kasiski, the distance between the beginnings of each sequence should be a multiple of the key length. We can then go back and look for sequences of length 5, length, 6, etc. If we can find a common factor across all these sets of duplicate sequences, that is a good guess for the key length. Doing this we get Table 4.5 of duplicates and their factors.

Note that all four of these duplicate patterns have a common factor of 6, leading us to surmise a key length of 6. This is not too far from the index of coincidence

Table 4.5 Repeated patterns in the example cryptogram

Pattern	Start	Offset	Difference	Factors
HMVG	3	159	156	2 3 4 6 12 13 26 39 52 78 156
IVWRYCG	15	111	96	2 3 4 6 8 12 16 24 32 48 96
VTTK	61	247	186	2 3 6 31 62 93 186
EZNWXFU	180	222	42	2 3 6 7 14 21 42

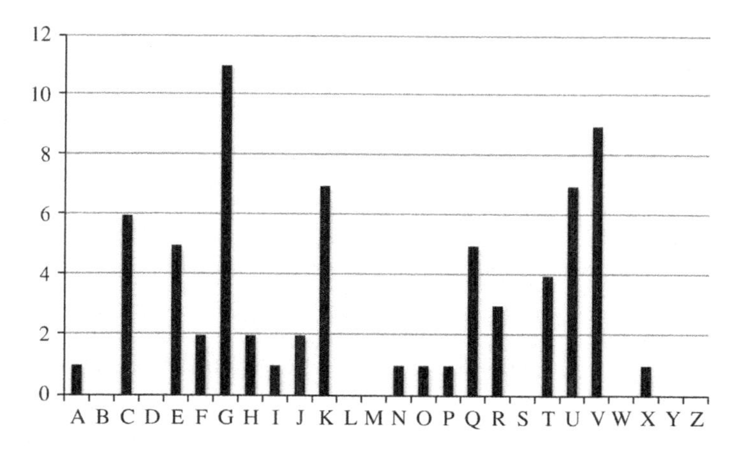

Fig. 4.2 Frequency chart for group 1

prediction of 4 or 5. If we go one step further and divide the cryptogram into six groups and compute the index of coincidence for each group, we get values of 0.0767, 0.0695, 0.0550, 0.0790, 0.0786, and 0.0562. This is the best set of values by far and certainly leads us to believe that the key is of length six.

If we have a key length of six, then each of our groups is a plaintext that has been enciphered using a monoalphabetic substitution with a shifted standard alphabet. Our next step is to try to find which alphabet was used for each group. The easiest way to do this is to create a frequency chart for each group and attempt to identify the letter E in each group. That will give us all the remaining letters in the alphabet. Figure 4.2 illustrates the frequencies of group 1.

Now this looks like a monoalphabetic substitution frequency chart. The E is plain to see as are the groups at JK, WX, and RST. If we then make a guess that G = e, then we'll have C = a, and C is the first letter of the key.

Continuing in this vein, we would compute the frequencies and draw the charts for the remaining five groups and discover that the keyword for this cryptogram is CRYPTO and the message is

> It is the strangest cipher I ever encountered, said Mr. Keen. The strangest I ever heard of. I have seen hundreds of ciphers, hundreds secret ciphers of the State Department, secret military ciphers, the elaborate oriental ciphers, symbols used in commercial transactions, ciphers used by criminals and every species of malefactor, and every one of them can be solved with time and patience and a little knowledge of the subject. But this one, he sat looking at it with eyes half closed, this one is too simple [5].

By the early part of the 20th century the old cryptographic algorithms were under increasingly sophisticated attack by cryptanalysts. The telegraph had increased the volume of communications in general, and enciphered traffic in particular. This meant that cryptanalysts had more ciphertext to examine and there were more opportunities for errors on the part of cryptographers and telegraphers. The increased volume of traffic meant that codes were at risk. The Babbage-Kasiski method and later Friedman's index of coincidence meant that the cryptanalysts finally had increasingly powerful weapons to use against polyalphabetic ciphers. What would come next?

References

1. Assarpour, Ali, and Kent D. Boklan. 2010. How we broke the union code (148 years too late). *Cryptologia* 34(3): 200–210.
2. Barker, Wayne G. 1978. *The history of codes and ciphers in the United States prior to World War I*, vol. 20. Laguna Hills, Calif: Aegean Park Press.
3. Bauer, Craig P. 2013. *Secret history: The story of cryptology*. Boca Raton, FL: CRC Press.
4. Boklan, Kent D. 2006. How I broke the confederate code (137 years too late). *Cryptologia* 30(4): 340–345.
5. Chambers, Robert W. 1906. *The tracer of lost persons*. New York: Appleton and Company.
6. Gaddy, David W. 1993. Internal struggle: The civil war. In *Masked dispatches: Cryptograms and cryptology in American history, 1775–1900*, ed. Ralph Edward Weber, pp. 105–120. Vol. 1. Ft. Meade, MD: Center for Cryptologic History, National Security Agency.
7. Kahn, David. 1967. *The codebreakers; The story of secret writing*. New York: Macmillan.
8. Singh, Simon. 1999. *The code book: The evolution of secrecy from Mary, Queen of Scots to quantum cryptography*. New York, NY: Doubleday.

Chapter 5
Crypto and the War to End All Wars: 1914–1918

Abstract The use of wireless telegraphy—radio—during World War I marked the advent of modern cryptology. For the first time, commanders were sending enciphered messages to front line troops and for the first time, the enemy had an enormous amount of ciphertext to work with. This spurred the development of more complicated codes and ciphers and eventually led to the development of machine cryptography. World War I is the first time that the Americans had a formal cryptanalytic organization. It is the beginning, in all the nations involved in the conflict, of the bureaucracy of secrecy. In the United States it marks the first appearance of the two founding fathers of modern American cryptology, Herbert O. Yardley and William F. Friedman. This chapter introduces Herbert Yardley and William Friedman and examines some of the cryptographic systems used during World War I.

5.1 The Americans Start from Behind

At the beginning of the 20th century there was no organized cryptologic effort in either of the military services of the United States—and there never had been. In all the conflicts in which the United States had been involved since it's founding, it had always had the occasional code, cipher, and cryptanalyst. And they had all been strictly ad hoc. In particular there had never been an official cryptanalytic organization in either the Army or the Navy. This was in sharp contrast to the Black Chambers of the European powers, which had been in existence since at least the 16th century.

The first real American cryptanalytic effort began in 1911 at the Army Signal School in Fort Leavenworth, Kansas. It was there that a few Army officers received initial training in cryptanalysis at a series of technical workshops. The students included Lt. Joseph Mauborgne who would one day head the Signal Corps and who, in 1914 published the first systematic solution of the Playfair cipher. Also trained at Leavenworth was Captain Parker Hitt who went on to write,

J. F. Dooley, *A Brief History of Cryptology and Cryptographic Algorithms*, SpringerBriefs in Computer Science, DOI: 10.1007/978-3-319-01628-3_5,

in 1916, *Manual for the Solution of Military Ciphers* that was to be the handbook for Army cryptanalysts for nearly a generation.

At America's entry into World War I in 1917, these two officers constituted about half of the trained cryptanalysts in the American military.

5.2 America Catches Up

In April 1917 Herbert O. Yardley (1889–1958) was 28 years old, a code clerk for the State Department in Washington, D.C. ambitious, and bored silly. Yardley had been with the State Department since 1912 and had pulled too many night shifts, waiting for diplomatic telegrams to come across his desk for encryption and decryption. At one point he decided to while away some time by trying to decode the personal correspondence between President Woodrow Wilson and his close aide Colonel House. Much to Yardley's surprise, it took him just a few hours to break the cryptosystem that Wilson and House were using [7]. Fascinated by the work of cryptology and appalled by how insecure many of the State Department cryptosystems were Yardley spent several months producing a 100-odd-page memorandum on the codes and ciphers then in use at State. Once war was declared, Yardley set about trying to get the Army to put him in charge of a cryptanalytic bureau. He finally convinced Major Ralph Van Deman of Military Intelligence and in June 1917 Yardley was commissioned a second lieutenant and placed in charge of Military Intelligence, Section 8—MI-8—the new cryptologic section—and the first official one the Army had ever created. What Yardley lacked in real cryptanalytic experience he made up for in energy and in innate organizational ability. Before the year was out MI-8 grew from Yardley and one clerk to six sub-sections, Instruction, Communications, Code and Cipher Compilation, Shorthand, Secret Inks, and Code and Cipher Solution and by the end of the war had 165 personnel. Yardley's second-in-command was Dr. (later Captain) John M. Manly, head of the English Department at the University of Chicago. Manly started in MI-8 as the chief of the Instruction section, but later became Yardley's best cryptanalyst. Manly brought with him several colleagues from Chicago including Dr. Edith Rickert, with whom Manly would write several textbooks and spend 14 years after the war creating the definitive set of volumes on Chaucer's *Canterbury Tales*.

MI-8 solved cryptograms from a number of nations, but focused its attentions on Germany, Mexico, and later, Japan. The high point of MI-8's cryptanalysis during the war was the case of the German spy, Pablo Waberski. Yardley tells the tale, suitably embellished, in his best-selling and controversial tell-all book, *The American Black Chamber* [7]. Waberski was arrested crossing the U.S.-Mexico border in January 1918. In Waberski's luggage was found a letter in cipher, which was forwarded on to MI-8 in Washington. After several MI-8 cryptanalysts failed to solve the cryptogram, Manly began working on it with Dr. Rickert assisting him. The accounts vary, but either after a brief two-days or a long several weeks Manly broke the cryptogram, which was a variation of a route transposition cipher. The message itself was a letter of introduction that plainly named Waberski as a German spy and laid out the sabotage that he was to attempt while in the United States. In August 1918, Yardley and Manly traveled to Fort Sam Houston in San Antonio, Texas where Manly testified on the exact nature of his cryptanalysis and the contents of the cryptogram. Waberski was convicted and sentenced to death; the only German spy given the death sentence during World War I. In the end, President Wilson commuted Waberski's sentence and he was released and sent back to Germany in 1923 [7].

Yardley was sent to England and France in August 1918 to establish closer relations with the cryptologic organizations there, leaving Manly in charge of MI-8. The English were very reticent about sharing anything with Yardley, and he was never given entrance into Room 40, the main Admiralty cryptologic organization (and the group that deciphered the Zimmermann Telegram that helped bring the United States into the war). The French were more cordial and Yardley met many of the cryptanalysts in their organization including Georges Painvin, the best French cryptanalyst of the war. The French, however, would not talk to Yardley about diplomatic codes and ciphers. After the Armistice, Yardley, now a major, was ordered to head the cryptologic section of the American delegation to the Versailles Peace Conference, and did not return to the United States until April 1919 at which point most of MI-8 had already been demobilized as the Army prepared for peace [3, pp. 354–355].

5.3 The A.E.F in France

While MI-8 in Washington focused on more strategic and diplomatic cryptologic systems, the American Expeditionary Force in France had it's own cryptologic organization that focused on tactical codes and ciphers. In the summer of 1917 as American forces were beginning to arrive in France, the cryptologic functions of the American Army were divided between Military Intelligence and the Army Signal Corps. The Radio Intelligence Section of Military Intelligence, designated as G.2 A.6 was organized under Major (later Colonel) Frank Moorman, one of the few Army officers trained in cryptanalysis. G.2 A.6 was primarily charged with code and cipher cryptanalysis, but also had sub-sections for traffic analysis, enemy telephone interception (via wiretaps), and monitoring of American communications to ensure security rules were followed. The Signal Corps had two sections devoted to codes and ciphers: the Code Compilation Section under Captain Howard Barnes, [3, p. 326] and the radio interception section that grabbed German cryptograms out of the air and passed them on to G.2 A.6. These organizations mirrored in many ways the cryptologic organizations of the British and French.

Many of the cryptologic personnel in France were trained by Yardley's organization in Washington and then shipped to American headquarters in Chaumont, to become either part of the Signal Corps, or Military Intelligence, section G.2 A.6. Among the cryptanalysts assigned to G.2 A.6 was Lieutenant William F.

Friedman, who arrived in France in July 1918. Friedman had trained cryptanalysts early in the war at Riverbank Laboratories before Yardley's organization was set up. Friedman was assigned at his own request to the code cryptanalytic section and spent the remaining five months of the war deciphering German *Satzbuch* and *Schlüsselheft* code messages. The Germans and the Allies both had decided that ciphers were too difficult to use near the front lines and so had reverted to 1-part and 2-part codes with anywhere from 800, to about 2,000 code groups for these *trench codes*. The Satzbuch codes were changed once a month and so the Americans had to break the codes quickly in order to be able to gain intelligence from the German secret messages. Friedman gained much experience with codes, something he had not had before, and went on to write the official monograph on *Field Codes Used by the German Army during the World War*, and also the history of the Code and Cipher Solving branch of G.2 A.6 [2, p. 69].

One of the first assignments of the Code Compilation Section of the Signal Corps was to create a trench code for the American Army. Barnes' organization had no experience with creating these types of codes, so they began with an obsolete British trench code and modified it for the American sector of the Western Front. The result was the American Trench Code of 1,600 codewords. It was a 1-part code and was designed to be *superenciphered*—the code message was enciphered using a monoalphabetic cipher—before any messages were sent. Because the Americans had no experience with this type of code before, Parker Hitt, then the chief of the Signal Corps for the A.E.F asked Lt. J. Rives Childs to see if he could recover the encipherment alphabet. If Childs could undo the superencipherment it would severely weaken the code. Childs sat down with 44 relatively short superenciphered messages in the American Trench Code and within 5 h had recovered the entire cipher alphabet.

Barnes scrapped the American Trench Code and proceeded to create one of the best series of trench codes in the war. The so-called *River Series* trench codes—all were named after American rivers—were 2-part codes with about 1,800 code words, nulls, and specific codewords for tactical use. The first code, *Potomac*, was released 24 June 1918 and Barnes' organization released a new code on the average of every two weeks for the rest of the war. In October, when the American 2nd Army was formed, a new series, the *Lake Series*, was begun and those codes were issued at the same rate as the River Series codes [3, p. 327].

5.4 Ciphers in the Great War: The Playfair

While all the combatants in World War I reverted to trench codes for much of their tactical communications, ciphers were not totally forgotten. In particular, the British used a field cipher as their tactical communications system for at least the first two years of the war, and the Germans used a complex field cipher for their high-level communications till the end of the war.

Sir Charles Wheatstone, the physicist, mathematician, and engineer, invented the British system, known as the Playfair cipher, in 1854. It acquired its name from Baron Lyon Playfair, who spent years popularizing the cipher and attempting to get the British government to adopt it. The British Army finally adopted the Playfair in the 1890s as their field cipher. It saw its first use during the Boer War (1899–1902) and was still used as the field cipher down to the company level during the first years of World War I [3, pp. 198–202, 1, pp. 166–178].

The Playfair cipher is a *digraphic substitution cipher* that encrypts two letters at a time. Every plaintext digraph is encrypted into a ciphertext digraph. It is based on a five by five Polybius square that uses a keyword to map 25 of the 26 letters of the Latin alphabet (I and J are either mapped together in a single cell, or J is just dropped). The keyword is dropped in row-by-row, deleting any repeated letters, and then the rest of the alphabet is filled into complete the square. For example, if the keyword is MONARCHY, then the Playfair square looks like Fig. 5.1.

Messages are enciphered according to the following rules:

1. The plaintext message is broken up into two-letter groups. Any double letters (like SS or LL) are broken up by inserting a null letter (like Q or X or Z) between the repeated letters. If the message has an odd number of letters, just add a null to the end.
2. Each two-letter group is enciphered separately.
3. If the two letters in a group are in the same row, then the group is enciphered by taking the letter immediately to the right of each letter in the group. So if the square in Fig. 5.1 is used and the plaintext pair is HY, then the ciphertext is YB. If you run off the right side of the square, just loop around to the beginning of the row.
4. If the two letters in a group are in the same column, then the group is enciphered by taking the letter immediately below each letter in the group. So in Fig. 5.1, if our plaintext is CL, then the ciphertext is EU. If you run off the bottom of the square, just loop around to the top of the column.
5. If the two letters are in different rows and columns then you "complete the rectangle" by first going across the row where the first letter is, to the column that contains the second letter and using the letter you find at the intersection as the cipher letter. Do the same thing for the second letter. So in Fig. 5.1 if our plaintext is MG, then the ciphertext is NE, in that order [6].

Deciphering is just the inverse of enciphering.

Say we want to send the message *flee, all is discovered* using a Playfair cipher with the keyword FRIEDMAN. Then the Playfair square will look like Fig. 5.2.

Fig. 5.1 Example of a Playfair cipher square

M	O	N	A	R
C	H	Y	B	D
E	F	G	I/J	K
L	P	Q	S	T
U	V	W	X	Z

Fig. 5.2 Playfair square using the keyword FRIEDMAN

F	R	I	E	D
M	A	N	B	C
G	H	K	L	O
P	Q	S	T	U
V	W	X	Y	Z

Then the first thing we do is divide up our plaintext into digraphs, making sure to break up any repeated letters with nulls

```
FL EX EA LX LI SD IS CO VE RE DX
```

We now use the rules above to encrypt each digraph separately

```
Plain:  FL EX EA LX LI SD IS CO VE RE DX
Cipher: EG IY RB KY KE UI NX OU YF ID IZ
```

And finally we break the ciphertext up into five-letter blocks for transmission

```
EGIYR BKYKE UINXO UYFID IZ
```

Cryptanalyzing a Playfair Cipher David Kahn gives an excellent description of the difficulties of solving a Playfair cipher

> In the first place, the cipher's being digraphic obliterates the single-letter characteristics—*e*, for example, is no longer identifiable as an entity. This undercuts the usual monographic methods of frequency analysis. Secondly, encipherment by digraphs halves the number of elements available for frequency analysis. A 100-letter text will have only 50 cipher digraphs. In the third place, and most important, the number of digraphs is far greater than the number of single letters, and consequently the linguistic characteristics spread over many more elements and so have much less opportunity to individualize themselves. There are 26 letters but 676 digraphs; the two most frequent English letters, *e* and *t*, average frequencies of 12 and 9 %; the two most frequent English digraphs, *th* and *he*, reach only 3.25 and 2.5 %. In other words, not only are there more units to choose among, the units are less sharply differentiated. The difficulties are doubly doubled. [3, pp. 201–202]

This is not to say that Playfair cipher messages are unsolvable; they are eminently solvable. For long Playfair ciphertexts, or when one has a large number of cipher messages, one can resort to digraph frequency analysis. Otherwise, luck, careful observation and a deep understanding of how the cipher works are the best methods. As mentioned earlier, U.S. Army Lt. Joseph Mauborgne was the first to publish a solution to a Playfair in 1914. In 1936, Alf Mongé published a detailed and easy-to-follow solution to a very short challenge Playfair [4]. And in her novel *Have His Carcase*, mystery writer Dorothy Sayers has her sleuth Lord Peter Wimsey walk through a very detailed and understandable solution of a Playfair cipher that solves the case [5, pp. 355–371].

5.5 Ciphers in the Great War: The ADFGVX Cipher

The most famous cipher of World War I was solved by the greatest cryptanalyst of the war. In the spring of 1918, both sides on the Western Front were exhausted, having fought to a standstill for nearly four years. The Germans knew that they had to crush the Allies soon, or they would run out of resources, both men and materiel. In preparation for their big spring offensives, the Germans changed their higher-level cipher system. This system was the one used to communicate at the division and corps level and above. The new system, called ADFGX appeared in early March, 1918 [3, p. 340]. It was different from any of the other cipher systems the Germans had used during the war.

ADFGX is what is known as a *fractionating cipher*. It is a substitution that produces digraphs as ciphertext, followed by a transposition where the digraphs are broken in two (the fractionating part) and then transposed.

It starts with a five by five Polybius square where a random mixed alphabet is inscribed in the square. The letters A, D, F, G, and X are used as both column and row headers of the square as in Table 5.1.

Encryption is now a three-step process. First, the message is read off one letter at a time and the corresponding row and column header becomes the digraph for that letter. Note that this operation will double the length of the message yielding Fig. 5.3.

Next, the digraphs are written out into a second table, row-by-row, one letter per column. The width of the table is the width of a pre-arranged keyword. If the keyword is GERMAN we get Fig. 5.4.

Finally, the fractionated table is sorted alphabetically by the keyword letters and the ciphertext is read off by columns Fig. 5.5.

```
AXDAA GDFDG GDAAX GXXGDX DDFFF DAGFG XDFFF A
```

Table 5.1 An ADFGX table with a mixed alphabet

	A	D	F	G	X
A	t	f	e	c	u
D	s	h	y	k	a
F	n	i	v	z	g
G	x	r	p	d	b
X	q	l	w	o	m

f	l	e	e	a	l	l	i	s	d	i	c	o	v	e	r	e	d
AD	XD	AF	AF	DX	XD	XD	FD	DA	GG	FD	AG	XG	FF	AF	GD	AF	GG

Fig. 5.3 First step of encryption using ADFGX

Fig. 5.4 Fractionated ciphertext

G	E	R	M	A	N
A	D	X	D	A	F
A	F	D	X	X	D
X	D	F	D	D	A
G	G	F	D	A	G
X	G	F	F	A	F
G	D	A	F	G	G

Fig. 5.5 Sorted ciphertext

A	E	G	M	N	R
A	D	A	D	F	X
X	F	A	X	D	D
D	D	X	D	A	F
A	G	G	D	G	F
A	G	X	F	F	F
G	D	G	F	G	A

The only way to solve an ADFGX cipher is to recover the sorted transposition key order. This is the problem that faced Georges Painvin on 21 March 1918 as the Germans launched their spring offensive.

Up to this point, less than three weeks after the ADFGX cipher had been introduced, there had not been enough traffic for Painvin to get a real handle on the cipher. But with the commencement of the offensive there was a jump in the number of messages transmitted and Painvin could really get to work.

Painvin noticed that there were messages in the pile of interceptions that had the same or very similar beginnings and a few with similar endings. He reasoned that this was because the plaintexts of these messages began with the same text and that the transpositions had moved the digraphs apart in a similar way. This was his key. Three weeks later, on 26 April he finally made a break in the initial group of interceptions and began to recover keys and break the cipher. His technique required a large number of messages and a subset of those with similar beginnings, so his technique would not work on all ADFGX messages and particularly he couldn't work with messages on days when there were few interceptions. Still, because the days immediately before an offensive saw an enormous increase in German traffic he was able to decrypt nearly 50 % of the messages sent.

Then just as he was hitting his stride and breaking more and more messages, the Germans changed the cipher on 1 June, adding an extra row and column to the Polybius square and an extra letter to the row and column headers. The cipher was now the ADFGVX cipher and each square now included all 26 letters of the alphabet and the ten decimal digits. Not too discouraged, Painvin worked for 26 h straight on the new messages and broke the updated cipher late in the day on 2 June [3, p. 345].

For a more detailed description of how Painvin solved the ADFGX cipher see [3, pp. 340–347]. For a description of a general solution of ADFGX, see [1, pp. 188–207].

5.6 A New Beginning

World War I marked the end of one phase in the history of cryptology. The volume of traffic that came as a result of the enormous armies that moved back and forth across Western Europe and their use of radio communication realistically marked the death knell for the lone cryptanalyst using paper and pencil to solve cryptograms one at a time. Radio allowed for the easy interception of messages and this increase in their number caused the cryptanalytic organizations in all the involved countries to grow enormously. Radio also added another dimension to cryptanalysis—traffic analysis. Traffic analysis allowed G.2 A.6 to tell the cryptanalysts where a message had come from and to whom it was addressed. This allowed the cryptanalyst to examine messages in more context than previously, giving him additional information and probable words to use. The enormous number of messages sent and received also caused a re-thinking of the methodology and process of cryptologic systems. Cipher systems in particular needed to be fast and easy to use, all the while providing an even higher level of security. The process of sending and receiving messages was found wanting in many areas as cipher clerks and telegraph operators made mistake after mistake both in enciphering and sending messages, giving more openings for the cryptanalysts to work their magic. Finally, the various intelligence bureaus and the general staffs at last came to the realization that cryptologic information was one of the most worthwhile and valuable forms of intelligence.

Speed, accuracy, simplicity, and increased security were desired going forward. The machines were on their way.

References

1. Bauer, Craig P. 2013. *Secret history: The story of cryptology*. Boca Raton, FL: CRC Press.
2. Clark, Ronald. 1977. *The man who broke purple*. Boston: Little, Brown and Company.
3. Kahn, David. 1967. *The codebreakers; The story of secret writing*. Hardcover. New York: Macmillan.
4. Monge, Alf. 1936. Solution of a playfair cipher. *Signal Corps Bulletin* 93.
5. Sayers, Dorothy. 1932. *Have his carcase: A Lord Peter Wimsey Mystery*. New York: Brewer, Warren & Putnam.
6. Singh, Simon. 1999. *The code book: The evolution of secrecy from Mary, Queen of Scots to quantum cryptography*. New York, NY: Hardcover.
7. Yardley, Herbert O. 1931. *The American Black Chamber*. Indianapolis: Bobbs-Merrill.

Chapter 6
The Interwar Period 1919–1939

Abstract In the period between the two World Wars Americans struggled with the morality and the cost of reading other people's mail. Herbert Yardley created his American Black Chamber and established for the first time that the United States should be in the position to protect itself and further its own interests with the use of permanent professional cryptographers and cryptanalysts. William Friedman, working in the Army, established the organization that would be the Army cryptologic backbone during the Second World War. Friedman and the team he put together during the 1930s would move American cryptology into the machine age in both cryptography and cryptanalysis. Despite Yardley's flaws and failure American would never again be without a cryptanalytic bureau. This chapter briefly examines the professional lives of Herbert Yardley and William Friedman and discusses their contributions to the growth of the American cryptologic infrastructure.

6.1 Herbert O. Yardley and the Cipher Bureau

When Captain Herbert Yardley returned from France in April 1919 it was to the prospect of demobilization and an almost certain return to the drab existence of a State Department code clerk. But nearly two years of being in charge of an exciting and important cryptanalytic organization had given Yardley more ambition than that. So he immediately began creating a plan for a "permanent organization for code and cipher investigation and attack." [9, p. 355] Yardley envisioned a joint State and War Department civilian organization that would be funded by both departments and would do all the training and cryptanalytic work for them. Being the consummate salesman that he was, Yardley got his funding. With a $100,000 budget–$40,000 from the State Department and $60,000 from the War Department—and about two dozen employees, mostly from MI-8 and the A.E.F organization Yardley set up shop in the fall of 1919. Yardley had originally wanted

J. F. Dooley, *A Brief History of Cryptology and Cryptographic Algorithms*, SpringerBriefs in Computer Science, DOI: 10.1007/978-3-319-01628-3_6,

50 employees but because the War Department never contributed its entire allotment of funds, he always had to manage with fewer. However, there was one fly in the ointment. According to the budget resolution for the State Department, no State Department funds were allowed to be expended in the District of Columbia. So Yardley was forced to move the entire organization to New York City and that's where the *Joint War-State Department Cipher Bureau* was housed for its entire existence. New York was where all the large telegraph companies had their headquarters and where most of the trans-Atlantic traffic passed through, making it an ideal location.

While the Cipher Bureau solved military, naval, attaché, and diplomatic cipher and code systems from many countries, its primary focus was on the diplomatic code systems of the great powers, especially Japan. Yardley, assisted by Frederick Livesey, a former MI-8 cryptanalyst, began working on the Japanese diplomatic code about the time that the Cipher Bureau was being formed in the summer of 1919. Yardley and Livesey worked almost continuously for five months attempting to find a way to break into the code. They were hampered by the fact that neither spoke Japanese, although Livesey taught himself Japanese over the course of a six-month period. The two cryptanalysts tried guess after guess but could make no headway into the code. Then, as Yardley relates, one night in December 1919,

> By now I had worked so long with these code telegrams that every telegram, every line, even every code word was indelibly printed in my brain. I could lie awake in bed and in the darkness make my investigations – trial and error, trial and error, over and over again. Finally one night I awakened at midnight, for I had retired early, and out of the darkness came the conviction that a certain series of two-letter codewords absolutely must equal *Airuando* (Ireland). Then other words danced before me in rapid succession: *dokuritsu* (independence), *Doitsu* (Germany), *owari* (stop). At last the great discovery! My heart stood still, and I dared not move. Was I dreaming? Was I awake? Was I losing my mind? A solution? At last – and after all these months!
>
> I slipped out of bed and in my eagerness, for I knew I was awake now, I almost fell down the stairs. With trembling fingers I spun the dial and opened the safe. I grabbed my file of papers and rapidly began to make notes. … I make a chart now in order to see how nearly correct I am, or at least to see in how many places the same meanings occur…
>
> Even this small chart convinces me that I am on the right track. For an hour I filled in these and other identifications until they had all been proved to my satisfaction.
>
> Of course, I have identified only part of the *kana* – that is, the alphabet. Most of the code is devoted to complete words, but these too will be easy enough once all the *kana* are properly filled in.
>
> The impossible had been accomplished! I felt a terrible mental letdown. I was very tired [12, pp. 268–271].

This was the first break into the Japanese diplomatic codes. This code, labeled Ja by Yardley, was the first of eleven different codes the Japanese used between the fall of 1919 and the spring of 1920. Others would be released at intervals throughout the remainder of the Cipher Bureau's existence.

The break into the Japanese diplomatic code was the high point of Herbert Yardley's cryptanalytic career. The break also led directly to the high point of the Cipher Bureau's achievements.

In November 1921 representatives from the United Kingdom, France, Japan, Italy, and the United States gathered in Washington for the Washington Naval Conference. After World War I diplomats from many countries were interested in limiting the growth and size of militaries around the world. The Washington Naval Conference's main goal was a treaty to limit the size of navies. To this end, the proposed Five Power Treaty tried to limit the ratios of tonnage of the navies of the participants. The Americans, supported by the British, had proposed a ratio of 10:10:6 for the tonnage of the navies of the United States, the United Kingdom, and Japan. The main sticking point of the conference was the Japanese insistence on a higher ratio for their navy.

Unbeknownst to the Japanese, Yardley and his Cipher Bureau were intercepting the telegrams between the Japanese negotiating team in Washington and the Foreign Ministry in Tokyo and decrypting them on a daily basis. A secure courier would ferry the translated decryptions from New York to the State Department in Washington every day. This meant that Secretary of State Charles Evans Hughes, who was representing the United States, was aware of the Japanese negotiating positions at all times. The most important decryption occurred on 28 November 1921 in a telegram from the Foreign Ministry in Tokyo to Baron Shidehara, the chief Japanese negotiator in Washington. Yardley gives us the entire telegram

> From Tokio
> To Washington.
> Conference No. 13. November 28, 1921.
> SECRET.
> Referring to your conference cablegram No. 74, we are of your opinion that it is necessary to avoid any clash with Great Britain and America, particularly America, in regard to the armament limitation question. You will to the utmost maintain a middle attitude and redouble your efforts to carry out our policy. In case of inevitable necessity you will work to establish your second proposal of 10 to 6.5. If, in spite of your utmost efforts, it becomes necessary in view of the situation and in the interest of general policy to fall back on your proposal No. 3, you will endeavor to limit the power of concentration and maneuver of the Pacific by a guarantee to reduce or at least to maintain the status quo of Pacific defenses and to make an adequate reservation which will make clear that [this is] our intention in agreeing to a 10 to 6 ratio.
> No. 4 is to be avoided as far as possible [12, p. 313].

The bottom line was that if the Americans would just wait and hold firm to their 10:10:6 ratio demand, the Japanese would agree. That was exactly what Hughes did and on 10 December the Japanese agreed to the American ratio.

The Washington Naval Conference was the high point of the Cipher Bureau's work. After their work peaked during the 1921–1922 Washington Naval Conference, the output of the Cipher Bureau decreased dramatically, along with its budget. At the beginning of the decade, during fiscal year 1920, the Cipher Bureau was allocated a budget of $100,000, although they never received that amount of money. By FY 1921, his budget was already down to $50,000 and four years later, it was $25,000: $15,000 from the State Department and $10,000 from the War Department, a level where it would stay for the remainder of the Cipher Bureau's existence [3, pp. 70–74].

By 1929, the Cipher Bureau was down to six people: Yardley, two other cryptanalysts (Ruth Wilson and Victor Weisskopf), and three clerks, including the future second Mrs. Yardley, Edna Ramsaier. Charles Mendelsohn, Yardley's partner in the Code Compilation Company—the Cipher Bureau's cover operation—worked part time for the Bureau. They were doing very little cryptanalysis because of the budget cuts and because of their inability to acquire any diplomatic cable or radio intercepts. The Radio Act of 1912 made it illegal to copy cablegrams, and the Radio Act of 1927 added a prohibition on radio interception as well [2, p. 18]. Because of this legislation, Yardley's friends at the telegraph companies were more and more reluctant to pass on any diplomatic cryptograms. What is more, the War Department was uninterested in Yardley's work because nearly all of the work the Cipher Bureau had done for all its existence was diplomatic traffic for the State Department. The War Department's interest was in the training of cryptanalysts for use in future wars, something that the Cipher Bureau had never done [1].

In July 1928, Signal Corps Major Owen S. Albright was placed in charge of the communications section of the Military Intelligence Division. Shortly thereafter, Albright's attention turned to Yardley's Cipher Bureau, and he was not pleased with what he saw. None of the four functions that Albright thought the Cipher Bureau should be performing for the Army—code and cipher compilation, code and cipher solution, radio interception, and training—were being done. The Cipher Bureau was performing code and cipher solution, but all its output was targeted at the State Department, not the War Department. Albright's conclusion was that all cryptologic activities for the Army should be centralized in one place, and that place should be the Signal Corps. By the beginning of 1929, Albright had decided that at least code and cipher compilation, code and cipher solution, training, and radio interception should be centralized in the Signal Corps for efficiency and to provide a single point of contact for the General Staff. In early 1929, Albright wrote a memorandum detailing his suggestions, and that memo began to work its way through the Army bureaucracy. Yardley became aware of the memorandum and Albright's intentions and may have begun thinking about moving the Cipher Bureau completely under the State Department's control [8, p. 88].

1929 brought a new President to the White House and a new Secretary of State into office. Because the State Department was providing $15,000 of the Cipher Bureau's budget each year, at some point the Secretary would have to be informed of its existence. In early May 1929 (no exact date has been found), Yardley shipped a batch of decoded Japanese diplomatic messages to the new Secretary of State, Henry Stimson [10, p. 97]. Yardley was no doubt trying to impress Stimson with the output and excellent work of the Cipher Bureau and trying to prepare him for Yardley's request to move the Cipher Bureau completely under the State Department. Yardley's revelations brought an unexpected reaction from Stimson. He completely disapproved of the existence of the Cipher Bureau, and he ordered all State Department funding (60 % of the total) be discontinued.

In the meantime, on 10 May 1929, Change #1 to Army Regulation 105-5 was approved, moving all Army cryptologic activities to the Signal Corps.

In June 1929, the State Department agreed to give the employees of the Cipher Bureau three months of severance beginning 1 July 1929. After that, the Bureau having no money would officially shut down. Their work, however, would cease immediately.

Herbert Yardley was at loose ends at the end of 1929. If he was not terribly surprised by the War Department's decision to transfer his Cipher Bureau, he seemed genuinely taken aback and puzzled by the abrupt withdrawal of State Department funds that ultimately closed his organization. For nearly a decade, the entire output of his Bureau had been diplomatic traffic of use to the State Department. The sudden closure of his operation, which meant that the State Department was now totally blind to foreign diplomatic messages, was a colossal mistake in his opinion. Of course, he was also suddenly unemployed at the beginning of the Great Depression.

By August 1930, Yardley was feeling the financial pinch. He wrote to a friend on 29 August that he was broke and having to sell off all his investment properties. He said, "I'm not certain at all what I shall do". Soon, though, he had an idea—one that changed his relationship with Friedman and the War Department forever.

On 20 December 1930, Yardley met with an editor at the Bobbs-Merrill publishing company to discuss a book detailing all his activities with MI-8 and the Cipher Bureau over the past thirteen years. The editor was excited about the idea, and a contract for the book was signed in early January 1931 [10, p. 105]. He immediately began writing the book, which would be called *The American Black Chamber*.

On 17 February 1931, Yardley signed a contract for three articles excerpted from his book with the Saturday Evening Post [10, p. 110]. The articles, titled, *Codes*, *Ciphers*, and *Secret Inks*, appeared on 4 April, 18 April, and 9 May 1931. Yardley also worked out the final contract arrangements with Bobbs-Merrill and was paid a $500 advance on the delivery of the completed manuscript on 23 February 1931. The book began to roll off the presses in May and was officially published on 1 June 1931.

It was an instant success with the public, becoming a best seller within weeks with sales of 17,931 copies in the U.S. and even more in Japan. It received generally favorable reviews—with some notable exceptions. Yardley's writing style was melodramatic and over-the-top, but he was a great storyteller. He over-emphasized nearly ever scene and took credit for nearly every success. The only other member of either MI-8 or the Cipher Bureau mentioned in the book is John Manly. Yardley hit the talk circuit and pushed his book at speeches across the country.

While the public ate up the stories of dramatic code breaks, spies, and exotic female secret agents, Yardley's former colleagues in MI-8 were not so pleased. Most of the cryptologic community in the United States thought that publication of *The American Black Chamber* was at best unwise, and at worst unethical and possibly treasonous. None of them liked Yardley's exaggerations and self-serving stories. Nor did they like the fact that in some cases he had conflated stories and played fast and loose with the facts. Particularly vitriolic in his condemnation of Yardley's behavior and his book was William Friedman. Friedman thought that Yardley had violated the oath he swore in the Army to keep his cryptologic work secret. Yardley insisted that he had written the book first, in order to feed his family, and second to send the message that by not having an active cryptanalytic bureau, the American government was leaving itself at an enormous disadvantage in world affairs. Friedman and Yardley, who had been friends from World War I all through the 1920s would never be friends again. Yardley, because of his book, would also never work in American cryptology again [10, p. 105, 8].

That is not to say Yardley wasn't busy. The government seized a second book, *Japanese Diplomatic Secrets*, before it could go into print. Yardley also had the distinction of having a federal law, Public Law 37, "For the Protection of Government Records" passed through Congress and signed by President Roosevelt in June 1933 to prevent the publication of *Japanese Diplomatic Secrets* [10, p. 162, 5]. Throughout the rest of the 1930s he wrote magazine articles and detective and spy fiction. His first novel *The Blonde Countess* was picked up by Hollywood and made into the movie *Rendezvous* starring William Powell and Rosalind Russell. In 1938 he was contracted by the Nationalist Chinese government to create a cryptanalytic bureau in China to read Japanese military codes and ciphers. Yardley spent two years in China, then came back to the United States and negotiated a contract to write about his experiences for the War Department. In 1941 he was employed by the Canadians to create their cryptanalytic bureau and left there in early 1942 only because the British would not work with the Canadians while Yardley was there. Yardley didn't know it at the time, but his Canadian work was the end of his cryptologic career. He tried to get work in the War Department's successor agency to his Cipher Bureau with no luck. He spent the World War II years in the Office of Price Administration, and after the war he built houses and wrote another bestseller, *The Education of a Poker Player*, in 1957. Herbert Yardley died of a stroke on 7 August 1958 [10, p. 236].

6.2 William Friedman and the Signal Intelligence Service

William Friedman returned to the United States after his tour of duty in France and was demobilized from the U.S. Army on 5 April 1919. After some soul searching and several rounds of letters, Friedman and his wife, Elizebeth Smith Friedman (they had married in 1917 before Friedman joined the Army) returned to Colonel Fabyan and Riverbank Laboratories. Curiously, Herbert Yardley had tried to get Friedman to come to work at the Cipher Bureau as it was being set up in the summer of 1919. From correspondence it appears that in July they were very close to an agreement for Friedman to join the Cipher Bureau at a salary of $3,000 along with employment for Elizebeth Friedman. But then suddenly the Friedmans ended up back at Riverbank.

Friedman really didn't like Riverbank. Fabyan was a bully and a braggart, and was always trying to insinuate himself into the Friedman's personal life. Friedman went back to Riverbank as head of the Cipher Department and continued to do work on request for the Government. It was also a productive time for Friedman as he published several monographs on cryptology under the Riverbank Publications imprint. The monographs included his most famous work Riverbank No. 22, *The Index of*

Coincidence and its Application to Cryptography, published in 1920. The index of coincidence is a metric that can be used to estimate the length of a key in a polyalphabetic substitution cipher [11]. It is more than that and the importance of the ideas behind the *Index of Coincidence* cannot be overemphasized. David Kahn wrote

> Before Friedman, cryptology eked out an existence as a study unto itself, as an isolated phenomenon, neither borrowing from nor contributing to other bodies of knowledge. Frequency counts, linguistic characteristics, and Kasiski examinations – all were peculiar and particular to cryptology. It dwelt a recluse in the world of science. Friedman led cryptology out of this lonely wilderness and into the broad rich domain of statistics. He connected cryptology to mathematics ... When Friedman subsumed cryptanalysis under statistics, he likewise flung wide the door to an armamentarium to which cryptology had never before had access. Its weapons – measures of central tendency and dispersion, of fit and skewness, of probability and sampling and significance – were ideally fashioned to deal with the statistical behavior of letters and words. Cryptanalysts, seizing them with alacrity, have wielded them with notable success ever since [9, pp. 383–384].

In late 1920, William Friedman finally broke loose from George Fabyan and the Riverbank Laboratories, and on 1 January 1921, he began a 6-month contract with the War Department as a cryptologist. In November 1921, he was hired as the Department's Chief Cryptanalyst, a post he still held at the beginning of 1929. With a single clerk, Friedman was the entire personnel of the Code and Cipher Section of the Signal Corps all through the 1920s [4, p. 19]. While Friedman was primarily charged with constructing codes and ciphers for the Army, he also put together the skeleton of a training regime, wrote the first version of his famous *Elements of Cryptanalysis*, and on occasion solved cryptograms for the War Department and other organizations [6].

Friedman also became well known within government circles during the 1920s. He was chosen as the U.S. technical advisor to the International Radiotelegraph Conference, held in Washington in November 1927, and was the technical advisor and Secretary of the U.S. delegation to the International Telegraph Conference, Brussels, Belgium in September 1928 [7]. By 1929, Friedman had solidified his role in the War Department, and when Major Owen S. Albright began to think about re-organizing the Army's cryptologic efforts, it was natural that he thought of Friedman and his organization in the Signal Corps.

At the beginning of 1929, William Friedman had been the sole cryptologist for the U.S. Army since 1921, and had a staff of exactly one clerk. That all changed dramatically when Army Regulation 105-5, Change #1 on 10 May 1929 officially brought all cryptologic work of the Army within the purview of the Signal Corps. On that day, the Signal Intelligence Service (SIS) came into being. It took the Army about eight more months, though, to put all the pieces together to really create the organization. On 13 January 1930 Friedman was authorized to hire four junior cryptanalysts for SIS. He was now ready to go.

Because of the special skills necessary for the SIS, it was not possible to find people on the current Civil Service rolls, and so Friedman was given permission to write his own requirements and look further afield. His searches bore fruit, and in March 1930 he was able to hire two clerks, Laurence Clark and Louise Nelson, followed in April by Frank Rowlett, Abraham Sinkov, and Solomon Kullback, his new junior cryptanalysts. Finding a fourth cryptanalyst with expertise in Japanese proved more difficult, and John Hurt was hired on 13 May as a cryptanalyst aide instead. This brought the strength of SIS up to seven, a number at which it would remain until fiscal year 1937 [4, p. 203].

Friedman immediately began a training regimen for his new junior cryptanalysts, using his own materials and a library of classic texts in cryptology he had acquired over the years. Their initial training focus was on breaking code and cipher systems, particularly those of Japan. Still without a radio interception service, and without Yardley's connections in the telegraph companies, Friedman and his students used old Japanese cryptograms from the Cipher Bureau files and some diplomatic intercepts provided by the Navy for their training. Circumstances in the 1930s would change the availability of training materials. Between the Japanese invasion of China, the Italian invasion of Ethiopia, the German acquisition of Austria and the Sudetenland, and other world events leading up to the Second World War, Friedman and his team would not lack for traffic on which to practice [4, p. 77].

Friedman's training regime kept his junior cryptanalysts busy for nearly two years during which he introduced two significant elements to the training. First was the study of cipher machines. The first electromechanical cipher machines began to be patented in 1919, less than a year after the end of the First World War. All through the 1920s new machines and improvements on existing machines had been introduced. Friedman had kept up with much of this work and passed that knowledge on to his students. Cipher machines will be covered in more detail in Chap. 7. The second significant change was the use of IBM accounting machines to improve efficiency in two areas, code compilation and cryptanalysis. Using the IBM machines gave the team a ten-fold improvement in the time required to create and print a 10,000-codeword field code. This was just the beginning of SIS' work on machine cryptography and cryptanalysis.

References

1. Albright, Major Owen S. 1929. Memorandum for ACoS, Signal Corps, ed. Signal Corps ACoS. Washington DC: NARA, DC, RG 457, Entry 9037 (SRH-038), Box 8.
2. Angevine, Robert G. 1992. Gentlemen do read each other's mail: American intelligence in the interwar era. *Intelligence and National Security* 7(2): 1–29.
3. Barker, Wayne G. 1979. *The history of codes and ciphers in the United States during the period between the World Wars, Part I. 1919–1929*. Cryptographic series, Vol. 22. Laguna Park, CA: Aegean Park Press.
4. Barker, Wayne G.: *The history of codes and ciphers in the United States during the period between the wars: Part II. 1930–1939*. Cryptographic series, Vol. 54. Laguna Park, CA: Aegean Park Press.
5. Bauer, Craig P. 2013. *Secret history: The story of cryptology*. Boca Raton, FL: CRC Press.
6. Callimahos, Lambros D. 1974. The Legendary William F. Friedman. *Cryptologic Spectrum* 4(1): 8–17.
7. Clark, Ronald. 1977. *The man who broke purple*. Boston: Little, Brown and Company.
8. Dooley, John F. 2013. 1929–1931: A transition period in U.S. cryptologic history. *Cryptologia* 37(1): 84–98.

9. Kahn, David. 1967. *The codebreakers: The story of secret writing*. New York: Macmillan (hardcover).
10. Kahn, David. 2004. *The reader of gentlemen's mail: Herbert O. Yardley and the birth of American codebreaking*. New Haven: Yale University Press (Vol. Book, Whole).
11. Singh, Simon. 1999. *The code book: The evolution of secrecy from Mary, Queen of Scots to quantum cryptography*. New York, NY: Doubleday.
12. Yardley, Herbert O. 1931. *The American black chamber*. Indianapolis: Bobbs-Merrill.

Chapter 7
The Coming of the Machines: 1918–1945

Abstract The volume of cipher traffic that was made possible by radio showed the need for vastly increased security, speed and accuracy in both enciphering and deciphering messages. The use of mechanical and electromechanical machines to do the encipherment was a logical outgrowth of this need. The first electromechanical rotor cipher machines began to appear right after World War I and the next three decades saw their steady improvement in both complexity and speed. The Enigma, the Typex and the M-134C/SIGABA were the epitome of these machines and the efforts to create and cryptanalyze them led us into the computer age. This chapter examines the history of cipher machines in the 20th century and looks in some detail at the cryptographic construction of the Enigma and the M-134C/SIGABA.

7.1 Early Cipher Machines

By the time World War I had ended all the nations involved in the conflict realized that they needed faster, more efficient, and more secure ways of enciphering messages in the field. They also needed much greater security at the division, corps, and army levels because so many important strategic messages were now transmitted via radio. That, combined with the fact that soldiers at or near the front lines were often lax about security protocols and often made mistakes, [4, p. 331] resulted in the officers in military intelligence looking for systems that could meet all these needs. They decided on machines.

Cipher machines had been proposed, if not widely used, as far back as Alberti and his cipher disk. Thomas Jefferson, Etienne Bazeries, Parker Hitt, and Joseph Mauborgne had all devised variations on the same device—the cipher cylinder. Mauborgne and Hitt combined to take an idea of Hitt's for a polyalphabetic cipher system based on the Bazeries system that used alphabets written on strips of cardboard that slid along in a frame and turn it into a cipher cylinder designated the

J. F. Dooley, *A Brief History of Cryptology and Cryptographic Algorithms*, SpringerBriefs in Computer Science, DOI: 10.1007/978-3-319-01628-3_7,

M-94 by the Army when it was adopted in 1922; it was used through World War II. An M-94 is pictured in Fig. 7.1. Hitt would later improve his original strip system and both the Army and the U.S. State Department adopted it as the M-138-A in the 1930s [4, p. 325].

7.2 The Rotor Makes its Appearance

Within five years of the end of World War I four different men in four different countries developed and patented devices that would generate polyalphabetic ciphertext. All the devices were electro-mechanical, all used standard typewriter keyboards, all were relatively small, and all used a new device that automatically allowed the machine to change alphabets whenever a plaintext letter was entered—the rotor.

An *electromechanical rotor* is a disk with 26 electrical contacts on either side. The disk is usually manufactured in two pieces so that the contacts on one side can be connected via wires to the contacts on the other side. The contacts are connected randomly from one side to the other so that when an electric current is passed through a contact on one side of the rotor, it appears at a different contact on the other side. The effect of the rotor is to create a monoalphabetic substitution cipher using a mixed alphabet. With just one rotor, a cipher machine would not be very secure. But what all rotor machines have in common is that once the rotor has been used to encipher a single plaintext letter, it is rotated one or more positions, presenting to the user a *different* substitution alphabet. If the user first types an A and gets a D as an output, and then types an A again, the second output could be an M, etc. Once every 26 letters the rotor gets back to the original alphabet, so we say that the *period* of the rotor is 26. This is still not very secure, being roughly equivalent to a Vigenère cipher. But if you put two rotors together things become much more interesting. The electrical current will then flow from an input contact on the first rotor and pass from an output contact in the first rotor to an input contact in the second rotor, and then take the output from the opposing contact on the second rotor, doing, in effect two substitutions. If the first rotor moves with every letter and the second rotor remains stationary, we still have just 26 alphabets. But, if when the first rotor has finished a complete rotation, the second rotor then advances one contact, then we have a different set of 26 alphabets to use. The period then becomes $26 * 26 = 26^2 = 676$ and the machine is using the equivalent of 676 mixed cipher alphabets. This is much more difficult to decrypt. Adding a third rotor brings the period to $26^3 = 17{,}576$, a fourth yields $26^4 = 456{,}976$, and a fifth $26^5 = 11{,}881{,}376$ alphabets. (Remember that there are $26! = 403{,}291{,}461{,}126{,}605{,}700{,}000{,}000{,}000$ possible mixed alphabets). Figure 7.2 shows an example of an electromechanical rotor.

Edward Hebern (1869–1952) from the United States was the first to develop an electromechanical cipher machine, called the Mark I in 1918. Hebern's original machine used standard electric typewriters and only a single rotor, but shortly after release he re-designed it so that it used up to five rotors. He started marketing his machines in 1921, primarily to the U.S. Army and Navy. Hebern was a better

Fig. 7.1 U.S. Army M-94 cipher device (National Cryptologic Museum)

inventor than a businessman though, and his first company went bankrupt in 1926. He tried again starting another company and began to get some business from the Navy until they abruptly canceled their contract in 1934 and Hebern was out of business again [4, p. 415–421].

Hugo Koch (1870–1928) of the Netherlands filed a patent application for an electromechanical rotor machine on 7 October 1919. Alas, that was it for Mr. Koch since he never formed a company and never marketed his device. Instead in 1927 he assigned his patent to an enterprising German, Arthur Scherbius.

Arthur Scherbius (1878–1929) was a German entrepreneur and engineer who patented his first rotor cipher machine in 1918. He called it the Enigma. He started a business to market the Enigma and others of his inventions, but struggled until 1926 when the resurgent German Navy decided to adopt a modified version of the Enigma. The German Army adopted the Enigma two years later and while Scherbius' company continued to struggle, a successor company founded in 1934 was a success.

The fourth man to develop a rotor machine was Arvid Damm, a Swede, who filed his patent in 1919 at almost the same time as Hugo Koch. Damm's claim

Fig. 7.2 Example of an electromechanical rotor

to cryptologic fame is due to the fact that the company he founded in 1915, AB Cryptograph, which later marketed his cipher devices, was the most successful cipher machine company. This was largely because of Boris Hagelin (1892–1983), a mechanical engineer who became the manager of the company in 1925. It was Hagelin who re-worked Damm's patent, removed the rotors and replaced them with a matrix of electrical contacts and a set of key wheels that used a set of pins to make contact with the matrix. Each key wheel used a different set of pins (and thus represented a different number of letters of the alphabet) and the period of the machine was the product of the number of pins on the key wheels. Hagelin called the machine the B-21 and a contract he signed with the Swedish Army in 1926 saved the company. Later Hagelin added a printing unit to the B-21 and developed the B-211. It only weighed 37 pounds, could operate at up to 200 letters per minute and wasn't much bigger than an electric typewriter. In 1934, following a request from the French Army for a cipher device that could fit in a uniform pocket and didn't use electricity, Hagelin designed the C-36. The C-36 was the first machine to use a "lug and pin" mechanism to rotate the rotors in an irregular fashion. A series of pins on the rotors would meet a lug on one of a number of horizontal bars in order to turn the rotor. Hagelin also added a device that printed its messages on an integrated paper tape.

Just before the beginning of World War II, Hagelin improved on the C-36 by adding movable lugs and another key rotor, resulting in the C-38. He varied the number of pins on the rotors and the number of letters on each rotor as well. One rotor had 26 pins and 26 letters, the next had 25 letters and pins (A–Z less W), then 23 (A–X less W), 21 (A–U), 19 (A–S), and finally 17 (A–Q). This gave the C-38 a total period of 26 * 25 * 23 * 21 * 19 * 17—101,405,850, resulting in a good degree of security.

In April 1940 as Germany was overrunning Norway, Denmark, the Netherlands, Belgium, and France, Hagelin and his wife, with the blueprints of the C-38 and two dismantled devices in their luggage, took a harrowing trip across Germany to take ship on one of the last ocean liners to leave Italy and sail for America. In the United States, William Friedman and the U.S. Army suggested some minor modifications and then approved the C-38 as the mid-level cipher machine for the U.S. Army with the designation Converter M-209. More than 140,000 M-209s were manufactured during the war and into the 1950s making Boris Hagelin the first person to be a cryptographic millionaire [4, p. 426–427] Fig. 7.3 illustrates an M-209.

7.3 How Does the Enigma Work?

The Enigma is an electromechanical cipher machine that uses rotors to create a set of polyalphabets to do both encryption and decryption. See Fig. 7.4 for an example.

The German Army, Navy, and Air Force all used the Enigma throughout World War II as their main mid- and high-level cipher machine. The Enigma is a self-inverse machine, so the same set-up and procedures are used for both encryption and

Fig. 7.3 Internals of the M-209 M-209B mechanical cipher machine http://en.wikipedia.org/wiki/File:M209B-IMG_0557.JPG free licence—Creative Commons Share-alike Attribution: Photograph by Rama, Wikimedia Commons, Cc-by-sa-2.0-fr

decryption. Originally, the Enigma was used with a set of three rotors in fixed positions, yielding an alphabet period of 17,576. Later the three rotors were allowed to be placed in any order so there are 6 possible sets of three rotors. This increases the period to $6 * 17{,}576 = 105{,}456$. The Germans then added two more rotors, so the operator was selecting three out of five rotors to be placed in any position, yielding $5 * 5 * 3 = 60$ positions and a period of $60 * 17{,}756 = 1{,}054{,}560$. By the end of the war, the German Navy was using an Enigma version that placed four rotors in the machine at a time for $26^4 = 456{,}976$ alphabets in 120 different positions for a period of 54,837,120.

In addition to the rotors, there is a fixed half-rotor, the reflector (*Umkehrwalze*) where 13 of the electrical contacts on one side of the rotor were connected to the other 13 contacts. This reflected the electrical signal and caused it to go back through the rotors along a different path. The reflector itself adds nothing to the encryption, but it does prevent any letter from encrypting to itself, which is a weakness in the machine [1].

Finally, in the military version of the Enigma there is a plugboard (*Steckerbrett*) that allows between 6 and 10 pairs of letters to be connected to each other. The plugboard adds about 150 trillion combinations of letters to the period. Each time a key is pressed the electrical signal runs from the keys, through the plugboard, through the rotors, then the reflector, then the rotors again, then the plugboard again, and finally to a set of lamps that indicates the ciphertext or plaintext letter. This is shown in Fig. 7.5.

Either two or three operators would be required to send or receive messages. Three operators were required because the Enigma doesn't print; all the output letters are displayed via lamps. So one operator would read the plaintext, the second would type and call out the cipher letter, and the third would write down the ciphertext that was then transmitted via Morse code.

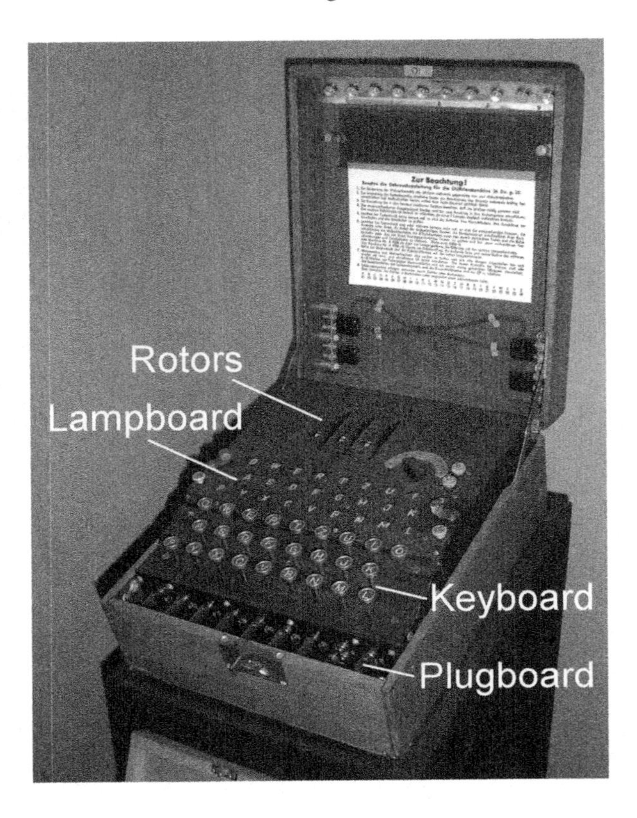

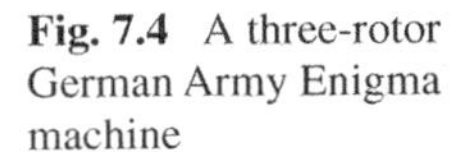

Fig. 7.4 A three-rotor German Army Enigma machine

Using the Enigma required that the machine be set up for each message using two keys, the *day key*, and the *message key*. The day key had five parts

1. the positions of the rotors (*Welzgelage*)
2. the plugboard settings (6–10 pairs) (*Steckerverbindungen*)
3. the turnover position on each rotor (*Ringstellung*). The ring is a notch in the rotor and is settable by the user. It is the position in the alphabet where the next rotor in line will advance one space.
4. the identification of the network (*Kenngruppen*). Each military branch had it's own network and set of keys.
5. the starting position of each rotor (*Grundstellung*) known to the British as the *indicator setting*; this is the letter on each rotor at which encryption will begin.

The *day keys* were distributed via courier once a month to all military units that used the Enigma.

The *message key* is the rotor setting for the current message. The procedure is as follows. The operator would set the day key on the machine. Then he would pick three random letters and encipher them. These are the first three letters sent in

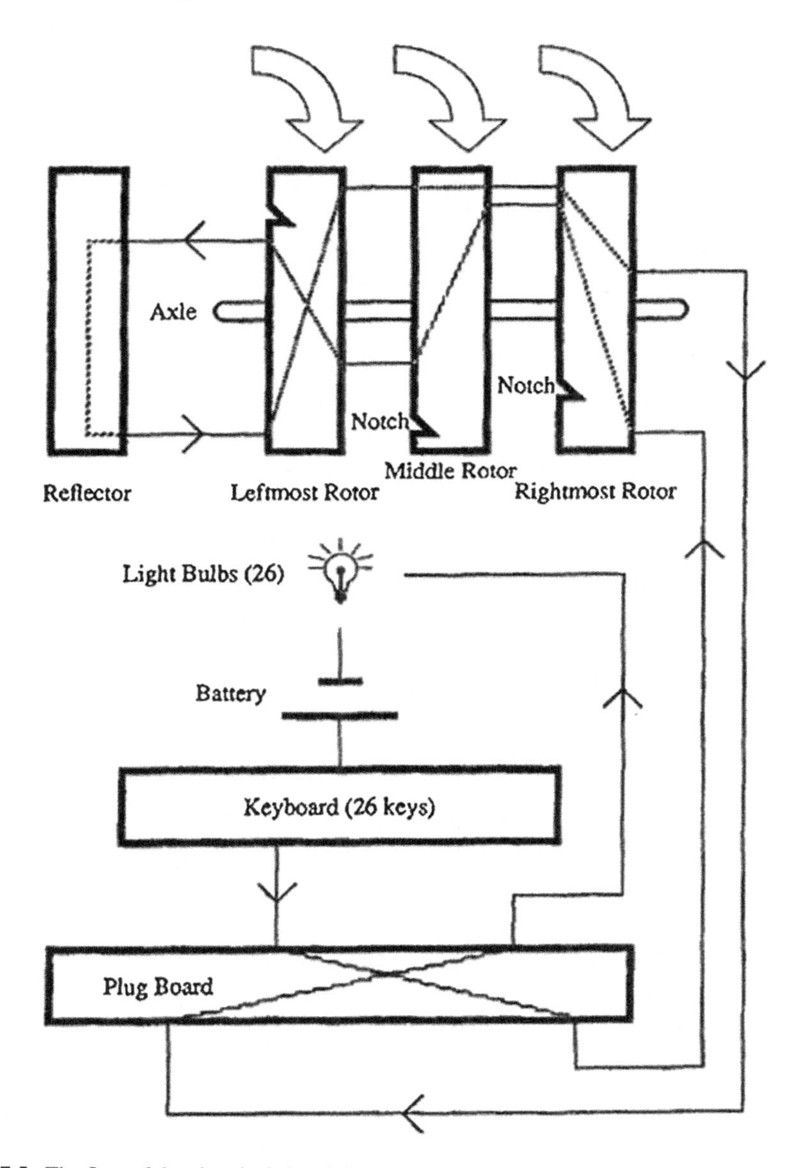

Fig. 7.5 The flow of the electrical signal through an Enigma [5]

the message. The operator would then reset the rotors to the three random letters. The machine is now set up to encrypt the message. On the receiving end, the operator would set the day key and then type in the first three letters of the message, recovering the message key. He would then reset the rotors to the message key and decipher the rest of the received cryptogram [1].

7.4 Solving the Enigma: Turing, Marian, and the Bombe

The German Navy and Army adopted and distributed the Enigma in 1926 and 1928 respectively. Given the complexity of the system and the number of possible alphabets, they were convinced it was an unbreakable cipher machine. And early on, they were right. The British and the French both acquired Enigmas and were unable to figure out a way to break into the ciphertext. William Friedman in the United States acquired an Enigma, studied it along with his junior cryptanalysts and also gave up. But one country had a very good reason to continue to try to break the Enigma until they succeeded. The Polish government knew that in the event of a new war, they were in the invasion path between Germany and Russia. The Germans were also anxious to reclaim Polish territory that had once been part of the German Empire, so the Poles spent as many resources as possible preparing to defend themselves. They created their own cipher bureau and recruited mathematicians to train in cryptanalysis. In the first class of cryptanalysts were Marian Rejewski, Henryk Zygalski, and Jerzy Rozycki. In September 1932 Rejewski began working on the German Army Enigma. Zygalski and Rozycki joined him in early 1933. By then Rejewski had had a breakthrough that enabled him to begin reading some messages. Rejewski's breakthrough was the result of two things. First, he had a brilliant mind and was able to cast the problem of key recovery in terms of permutation theory. He realized that he could separate the problem of the plugboard from the rotor behavior and that because of this behavior he could create "chains" of letters that would lead him to the key letters; these chains were permutation cycles. Rejewski's second piece of luck was that a German traitor, Hans Thilo Schmidt, was selling the day keys of the Enigma to the French. The French had given up on Enigma, but they were willing to pass the data on to the Poles. So by early 1933 the Poles were able to read an increasing number of German Army Enigma messages.

Then the Germans changed the playing field. First, in September 1938 they changed the indicator settings (the first part of the message that identified the message key). This made the Poles "chains" useless. Then in December 1938 they added two more rotors to the Army Enigma, increasing the amount of work tenfold. Finally, they also increased the number of pairs of letters connected by the plugboard from 6 to 10. At this point the Poles were out of resources and nearly out of time. On 24 July 1939, just five weeks before the start of the war, the Poles met with their French and British counterparts outside Warsaw and gave them everything they knew about the Enigma. The French and the British in particular started work on deciphering Enigma immediately [1].

On 4 September 1939, one day after the British declaration of war, Alan Turing arrived at Bletchley Park to begin working for the Government Code and Cipher School. Turing immediately focused on Naval Enigma [3].

By November 1939 Turing had an idea. The Poles had been defeated because their techniques attempted to find the rotor order, the *Ringstellung* or *ring settings* and the *Grundstellung* or *initial rotor positions* based on the indicator system that the Germans were using. So as soon as the Germans changed the indicator system, the Poles had to start all over again. Instead of trying to identify positive things, Turing decided to try to eliminate as many wrong answers as possible [5]. This would then reduce the number of rotor positions, ring and rotor settings they would have to try manually. He designed a machine called a *bombe* which took advantage of *cribs*—probable words in ciphertext—to find rotor settings, rotor order, and the plugboard settings by looking for mistakes and throwing them out. Turing's bombe (see Fig. 7.6) checked whether, with the current rotor order, the current rotor position, and any plugboard swapping, the crib and ciphertext could be transformed into each other. This was the break the British needed. The bombe didn't solve all the problems of Enigma and the British had a long 10-month period of darkness when the German Navy switched from a three-rotor to a four-rotor Enigma. But Turing's idea (the first of many over the next several years) was the first giant step in breaking Enigma [3].

7.5 SIGABA: Friedman and Rowlett's Triumph

The SIGABA with its curious history, began life as a design by William Friedman that was implemented as the Army Converter M-134. Friedman was trying to improve the security of rotor-based cipher machines by attempting to avoid the single stepping behavior of rotors in machines like the Enigma. Friedman reasoned that if the rotors advanced irregularly according to a separate key that it would be much more difficult to predict which alphabets were being used. He implemented this idea by integrating a paper tape reader into the M-134. The key that was punched into the paper tape controlled the stepping of the cipher rotors in the device. At one point Friedman asked Frank Rowlett to create a series of key materials—paper tapes with the keys on them. Rowlett had a great deal of difficulty with this chore because the procedure that Friedman had outlined was cumbersome and time consuming. Instead, Rowlett came up with an electromechanical way to generate the key stream randomly that didn't require creating any key materials a priori [6]. Once Friedman was convinced of the efficacy of Rowlett's idea, they created an add-on device, called the M-229 (also called the SIGGOO), to attach to the handful of M-134 s that had already been distributed to the field. Unfortunately, Friedman could not get the funds to develop an integrated device that combined the M-134 and M-229. He did, however, tell Lt. Joseph Wenger of the U.S. Navy about the integrated device. When Wegner passed this information on to Captain Laurence Safford, the head of the Navy's cryptanalytic group, OP-20-G, Safford was excited and proceeded to develop the device. In 1941 the Army and Navy finally got together on the device and completed development together. The Army called it the M-134C and the Navy the CSP-888, but the common name was SIGABA [2].

Fig. 7.6 The Turing bombe (from http://commons.wikimedia.org/wiki/File:Bletchley_Park_IMG_3606.JPG)

Friedman holds the patent for the original M-134,[1] and it is Safford and Sieler who hold the patent for the modified M-134C/SIGABA.[2] Finally, there is a second patent for an integrated M-134 device with a new set of rotors for controlling the key stream using "cam wheels" awarded to solely to Friedman on 10 October 2000, but filed on 23 October 1936.[3]

7.6 How Does the SIGABA Work?

SIGABA is a multi-rotor electromechanical cipher machine. It uses fifteen rotors: five cipher rotors, five control rotors, and five index rotors. The cipher rotors and control rotors are identical and interchangeable 26 contact rotors. They are inscribed with the letters of the alphabet on the outside ring. Also, the left and right sides of these rotors are identical so it is possible to insert the rotors into the machine backwards. The five index rotors only have 10 contacts each and are inscribed with the numbers 10–59 in sequence. So index rotor 1 has the numbers 10–19, rotor 2 has 20–29, etc. Unlike the Enigma, there is no reflector at the end of the cipher rotors.

---

[1] Patent 6,097,812, granted 1 August 2000.

[2] Patent 6,175,625, granted 16 January 2001.

[3] Patent 6,130,946, granted 10 October 2000.

When a key is pressed an electrical signal passes through a contact in the cipher rotors and the resulting output signal is the ciphertext letter. One or more of the cipher rotors then rotates, depending on the outputs of the control and index rotor groups.

The control rotors receive four signals and output up to four signals that are collected into ten groups that become the inputs to the index rotors. Of the five control rotors, the two outer rotors do not rotate, but the inner three rotate in exactly the same way as a three rotor Enigma. The ten groups connect the output contacts using logical OR to generate the signal in the following manner

1: A
2: B
3: C
4: D, E
5: F, G, H
6: I, J, K
7: L, M, N, O
8: P, Q, R, S, T
9: U, V, W, X, Y, Z
0: is grounded.

The index rotors receive the ten signals and route them through the five rotors. The index rotors do not rotate and their outputs are logically OR'ed by pairs. It is the output signals from the index rotors that cause the cipher rotors to rotate. At least one cipher rotor and at most four will rotate after every key press. This irregular stepping of the rotors is the key to SIGABA's security because it eliminates the predictable succession of cipher alphabets that machines like the Enigma produce. Once you know the rotor wiring of an Enigma you can predict the next alphabets. That is much more difficult to do with a SIGABA. A SIGABA is shown in Fig. 7.7 [4].

Fig. 7.7 SIGABA machine (National Cryptologic Museum)

References

1. Budiansky, Stephen. 2000. *Battle of wits: The complete story of codebreaking in World War II*. New York: Free Press.
2. Clark, Ronald. 1977. *The man who broke purple*. Boston: Little, Brown and Company.
3. Hodges, Andrew. 1983. *Alan turing: The Enigma*. New York: Simon and Schuster.
4. Kahn, David. 1967. *The codebreakers; The story of secret writing*. New York: Macmillan.
5. Miller, A. Ray. 1996. The cryptographic mathematics of Enigma, ed. Center for cryptologic history. Fort George G. Meade, Md.: National Security Agency.
6. Rowlett, Frank. R. 1998. *The story of magic: Memoirs of an American cryptologic pioneer*. Laguna Hills, CA: Aegean Park Press.

Chapter 8
The Machines Take Over: Computer Cryptography

Abstract Modern cryptology rests on the shoulders of three men of rare talents. William Friedman, Lester Hill and Claude Shannon moved cryptology from an esoteric, mystical, strictly linguistic realm into the world of mathematics and statistics. Once Friedman, Hill, and Shannon placed cryptology on firm mathematical ground, other mathematicians and computer scientists developed the new algorithms to do digital encryption in the computer age. Despite some controversial flaws, the U.S. Federal Data Encryption Standard (DES) was the most widely used computer encryption algorithm in the 20th century. In 2001 a much stronger algorithm, the Advanced Encryption Standard (AES) that was vetted by a new burgeoning public cryptologic community, replaced it. This chapter introduces Hill and Shannon and explores the details of the DES and the AES.

8.1 The Shoulders of Giants

Modern cryptology rests on the shoulders of three giants of the 20th century. We've already talked about William F. Friedman and how his theoretical work, particularly the *Index of Coincidence*, brought statistics to cryptanalysis. Two other mathematicians made even more impressive impacts on cryptology in significantly different ways.

Lester S. Hill (1890–1961) was a mathematician who spent most of his career at Hunter College in New York City. In the June/July 1929 issue of *The American Mathematical Monthly* he published a paper titled *Cryptography in an Algebraic Alphabet* that marched cryptography a long way down the road towards being a mathematical discipline [7]. Hill's paper and its sequel in 1931 [8] were the first journal articles to apply abstract algebra to cryptography [9]. The substance of his paper was a new system of polygraphic encryption and decryption that used

J. F. Dooley, *A Brief History of Cryptology and Cryptographic Algorithms*, SpringerBriefs in Computer Science, DOI: 10.1007/978-3-319-01628-3_8,

invertible square matrices as the key elements and did all the arithmetic modulo 26. This is now known generally as matrix encryption, or the Hill cipher [3, p. 227]. The fundamental idea is to convert the letters of a message into numbers in the range 0 through 25 and to apply an invertible $n \times n$ square matrix to the numbers to create the ciphertext. The beauty of the system is that you can use as many of the letters of the message as you like and encrypt them all at once—a true polygraphic system. The system works by picking a size for the polygraphs, say 2. Then the user creates an invertible 2×2 matrix, M. The digraph letters are arranged as a two-row column vector (a 2×1 matrix) L and multiplying L by M creates the ciphertext. This looks like $M \cdot L = C$ where the $\cdot$ denotes matrix multiplication. Decryption just takes C and multiplies it by M^{-1} as in $M^{-1} \cdot C = L$. This system is easy to use but provides very good security. More importantly, Hill took another giant step in applying the tools of mathematics to cryptography.

The other mathematician we will discuss had the most significant and important impact on cryptology of the group. Claude Elwood Shannon (1916—2001) was both a mathematician and an electrical engineer and received his Ph.D. from M.I.T. in 1940. Two years earlier, his master's thesis was the first published work that linked Boolean algebra with electronic circuits—the basis of all modern computer arithmetic. In 1941 he joined the staff of Bell Telephone Laboratories and was soon working on communications and secrecy systems under contract from the War Department. In 1948 he was finally able to publish his work on communications systems as *A Mathematical Theory of Communication* [11], the foundational paper in information theory. In 1949 he followed with another seminal paper, *Mathematical Theory of Secrecy Systems* [12]. What Friedman had started and Hill continued, Shannon completed. In 60 dense pages *Secrecy Systems* placed cryptology on a firm mathematical foundation and provided the vocabulary and the theoretical basis for all the new cryptographic algorithms that would be developed over the next half-century. Shannon explored concepts like message entropy, language redundancy, perfect secrecy, what it means for a cipher system to be computationally secure, the unicity distance of a cipher system, the twin concepts of diffusion and confusion in cryptologic systems, product ciphers, and substitution-permutation networks.

Important for our discussion of computer algorithms are the concepts of *diffusion* and *confusion*. In general parlance, diffusion means spreading something widely across an area. A definition aptly used in Shannon's work. In Shannon's systems, messages are reduced to representations as numbers that are *binary digits* (bits) in a machine. A *secrecy system* is an algorithm that transforms a sequence of message bits into a different sequence of message bits. The idea of diffusion is to create a transformation that distributes the influence of each plaintext bit across a large number of ciphertext bits [3, p. 337]. Ideally the diffusion occurs across the entire ciphertext output. This is known as an *avalanche effect* because the effect of a single bit change is cascaded across many ciphertext bits. In a cipher, using transposition creates system diffusion. In diffusion the emphasis is on the relationship between the plaintext and the ciphertext. *Confusion* is the process of making the relationship between the plaintext and the ciphertext as complex as possible.

A cipher system does this via substitution [3, p. 337]. This complicates the transformation from plaintext to ciphertext, making the cryptanalyst's work much more difficult. In confusion the relationship is between the key bits and the ciphertext (a change in the key bits will change ciphertext bits). Shannon combined these two ideas into a *substitution-permutation network* (*S–P*) that uses diffusion and confusion to complicate the cipher. He also suggested that executing an S–P network a number of times—a *product cipher*—will also make the system that much more resistant to cryptanalysis.

8.2 Modern Computer Cipher Algorithms: The DES

Horst Feistel (1915–1990) struggled for many years to be allowed to do the cryptologic research he really wanted to do. But working for the government and government contractors made it difficult. When he finally started work at IBM's T.J. Watson Research Center in Yorktown, NY in the early 1970s he was finally able to do his cryptologic research. The result was a system called *Lucifer*. Lucifer was a very secure computer-based cipher system that IBM marketed and sold within the United States and—in a weakened version—abroad [6]. This was in response to the increasing amount of business being done via computer and the increasing number of financial transactions being handled across networks. Then, in 1973, the National Bureau of Standards put out a call for cryptographic algorithms that would be a federal standard and would be used to encrypt unclassified government data. It was clear that any algorithm that was a federal standard would also become very popular in the business world, so IBM submitted Lucifer as a candidate. It turned out that Lucifer was the only acceptable algorithm and it was adopted as Federal Information Processing Standard 46 (FIPS-46) on 15 July 1977 and renamed the federal Data Encryption Standard or DES [1].

8.2.1 How Does the DES Work?

The DES is a *symmetric block cipher algorithm*. It uses a single key to both encrypt and decrypt data (the symmetric part). It operates on data in 64-bit blocks (eight characters at a time), using a 56-bit key. It passes each block through the heart of the algorithm—a *round*—16 times before outputting the result as ciphertext. Each round breaks the 64-bit block into two 32-bit halves and then implements a substitution-permutation network using part of the key, called a sub-key, to produce an intermediate ciphertext that is then passed back again for the next round. Figure 8.1 diagrams the data flow of a round [1].

In more detail, the 64-bit input to DES is put through an *initial permutation* (IP) that rearranges the bits. The 64-bits are then divided into two halves, Left and Right and put through a round. In a round, nothing is done to the Right half.

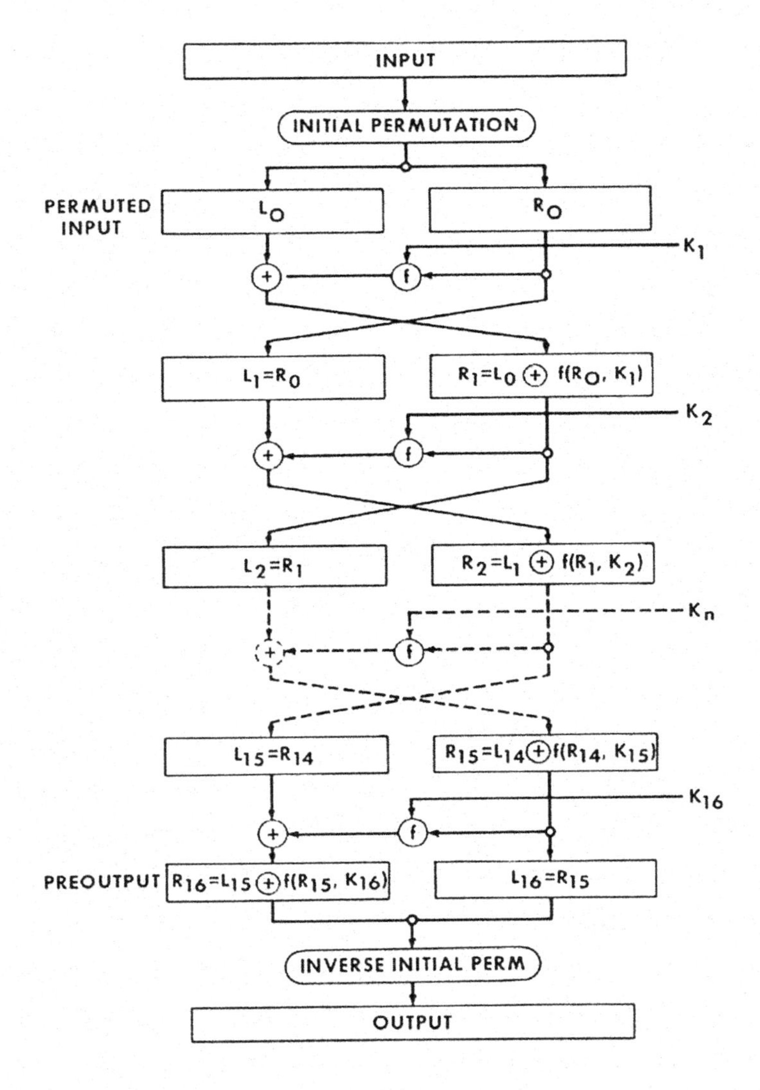

Fig. 8.1 Sixteen rounds of DES [1]

It becomes the Left half of the next round. The Right 32-bits are first put through a mixing function *f(Right, Key)* where Key is a sub-key generated by the *key scheduler*. The output of the function f() is exclusive-or'ed with the Left half. The result of this operation becomes the Right half input to the next round and the original Right half is the Left input to the next round. After the sixteenth round, the 64-bit output is put through a permutation that is the inverse of the initial permutation above. The resulting output is the 64-bit ciphertext.

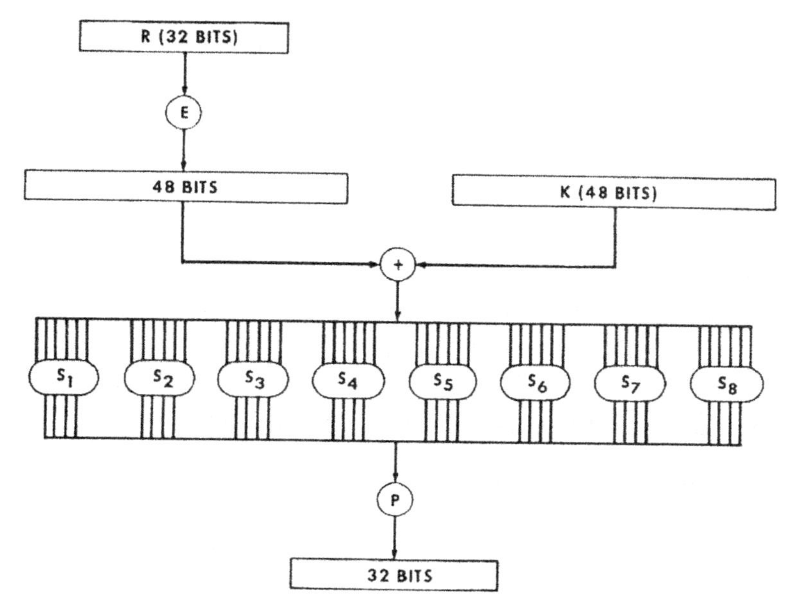

Fig. 8.2 Internals of the f() function [1]

8.2.2 The f() Function

The f() function takes as input the 32-bit right half of the input and a 48-bit sub-key generated by the key scheduler. This is illustrated in Fig. 8.2.

The first thing the f() function wants to do is XOR the right half with the sub-key. However, the generated sub-key is 48-bits and the right half of the data is only 32-bits. So the data must first go through the expansion block E. E performs a transformation that changes the right half into a 48-bit output using the expansion table in Fig. 8.3.

The 48-bit output of the exclusive or is broken up into 8 groups of 6 bits each and these 6-bit quantities are use as indexes into the substitution or S-boxes to select a four-bit output quantity. Block S_1 in Fig. 8.4 illustrates this selection.

The four-bit values from the 8 S-boxes are then combined and permuted one last time to make the 32-bit output of the function.

8.2.3 The Key Scheduler

The 56-bit DES key is broken up via the key scheduler into 48-bit sub-keys and a different sub-key is used for each round. The diagram for the key scheduler is in Fig. 8.5.

Fig. 8.3 E bit expansion table

E BIT-SELECTION TABLE

32	1	2	3	4	5
4	5	6	7	8	9
8	9	10	11	12	13
12	13	14	15	16	17
16	17	18	19	20	21
20	21	22	23	24	25
24	25	26	27	28	29
28	29	30	31	32	1

The original key is permuted and then broken up into two 28-bit halves. These halves are left shifted by an amount that depends on which round the key is destined for. The shift, though, is always either 1 or 2. The two 28-bit halves are then recombined, permuted, and 48-bits are selected for the round key. From the introduction of DES, ciphers that have this particular design of round are said to have *Feistel cipher structures* or *Feistel architectures*.

While the DES looks complicated, note that the only operations that are performed are XOR (exclusive or), bit shifting, and permutation of bits, all very simple operations in hardware. This allows DES to be fast.

8.2.4 Discussion of DES

With multiple substitutions and transpositions (disguised as permutations), the DES does a very good job of implementing Shannon's confusion and diffusion. It was not without controversy, though. Two particular areas stand out.

First, the key is too short [5, 10]. A 56-bit key only yields a key space of 2^{56} possible keys. This is only about 10^{18} or a quintillion keys, about half of which would need to be tried before the correct key was found to decrypt a message using brute force. Now this is not a small number, but with 201x computers we

S_1

Column Number

Row No.	0	1	2	3	4	5	6	7	8	9	10	11	12	13	14	15
0	14	4	13	1	2	15	11	8	3	10	6	12	5	9	0	7
1	0	15	7	4	14	2	13	1	10	6	12	11	9	5	3	8
2	4	1	14	8	13	6	2	11	15	12	9	7	3	10	5	0
3	15	12	8	2	4	9	1	7	5	11	3	14	10	0	6	13

Fig. 8.4 Substitution box S_1

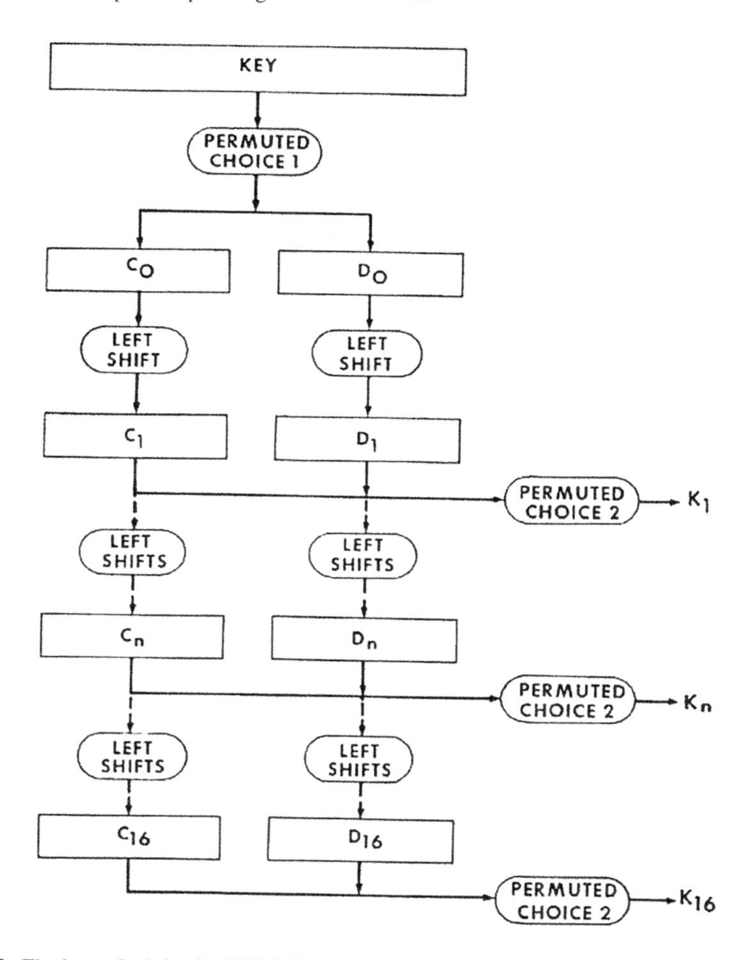

Fig. 8.5 The key scheduler for DES [1]

are talking less than a day to break a DES key. Even in the 1970s it was estimated that one could spend about $20 million dollars and create a special purpose machine that would break DES. In 1997 a network of thousands of computers on the Internet broke a DES key in a little over a month's time. And a year later, a special purpose computer built by the Electronic Frontier Foundation for less than $250,000 broke a DES key in less than three days [3, p. 385]. If a not-for-profit civil liberties organization can break a DES key in that short a time, surely a well-funded corporation or government can do it in less.

Why was the key so short? The original Lucifer key lengths were 64-bits and 128-bits, so why was the key shortened for DES? The prevailing theory at the time was that the NSA had requested the shorter key because their computing technology could break a 56-bit key in short order, but not anything larger.

The second piece of controversy is that at its introduction there was much complaint and discussion about the design of the substitution boxes of the DES. IBM and the NBS were closed-mouthed about how the particular values in each of the eight S-boxes were chosen and why [5, 10]. Again, suspicion fell on the NSA. This time the suspicion was that the design afforded the NSA a back door into the cipher. None of these accusations have been proven, and DES has stood up to heavy use for a quarter of a century. But by the mid-1990s it was beginning to show its age. Moore's law was making it more and more likely that cheap systems for breaking the DES would be available soon. So the National Institute of Science and Technology (the successor to the NBS) decided it was time for a new algorithm.

8.3 The Advanced Encryption Standard

In 1997 NIST sent out a call for potential successors for the DES. The climate was much different than in the early 1970s; by the 1990s there was a flourishing international community of researchers and practitioners in cryptology. Fifteen candidates were accepted and presented their algorithms at a NIST conference in 1998. By August 1999 the list was down to the top five candidates, RC6 from RSA, Inc in the U.S., MARS from IBM, Twofish from Counterpane in the U.S., Serpent from an English/Israeli/Danish group, and Rijndael from a group in Belgium. At this point all five algorithms were published and the international community was challenged to evaluate them and look for weaknesses. NIST also did its own evaluations.

In August 2000 Rijndael was chosen as the next standard and the new Advanced Encryption Standard (FIPS-197) was published in November 2001 [2].

AES is a symmetric key block cipher, just like DES. It uses a 128-bit input block, and gives the user three choices for key sizes, 128-bits, 192-bits, and 256-bits. The number of rounds varies depending on the key size. AES-128 uses 10 rounds, AES-192 uses 12 rounds, and AES-256 uses 14 rounds. The key data structure in AES is called *The State*. It is a 4×4 matrix of bytes (so 16 bytes * 8 bits/byte = 128-bits) that is acted upon by the algorithm to produce a 128-bit output. The basic algorithm for AES looks like Fig. 8.6.

In Fig. 8.6 N_b is the number of bytes in the input data block, and N_r is the number of rounds. Note that each round is basically four steps, *SubBytes*, *ShiftRows*, *MixColumns*, and *AddRoundKey*. The final round (outside the for loop) skips the MixColumns step.

Figures 8.7, 8.8, 8.9 and 8.10 illustrate each of these steps.

Notice that AES is not a Feistel architecture because it does not separate the input block into two halves; instead it is an iterative cipher, operating on the entire block in every round. It also is not invertible as written. To do decryption, you must apply the round structure in reverse. As with DES, AES provides

```
Cipher(byte in[4*Nb], byte out[4*Nb], word w[Nb*(Nr+1)])
begin
   byte  state[4,Nb]

   state = in

   AddRoundKey(state, w[0, Nb-1])                // See Sec. 5.1.4

   for round = 1 step 1 to Nr–1
      SubBytes(state)                            // See Sec. 5.1.1
      ShiftRows(state)                           // See Sec. 5.1.2
      MixColumns(state)                          // See Sec. 5.1.3
      AddRoundKey(state, w[round*Nb, (round+1)*Nb-1])
   end for

   SubBytes(state)
   ShiftRows(state)
   AddRoundKey(state, w[Nr*Nb, (Nr+1)*Nb-1])

   out = state
end
```

Fig. 8.6 The basic AES algorithm [2]

a key scheduler to create sub-keys, one for each round. The key scheduler is in Fig. 8.11.

Rijndael is designed to be easy to implement on architectures from 8-bit through at least 64-bit. Its operations can either be pre-computed or done using very simple operations. SubBytes just needs a table of 256 entries. ShiftRows is just simple byte shifting. MixColumns can also be implemented as a table look-up, and AddRoundKey just uses XOR.

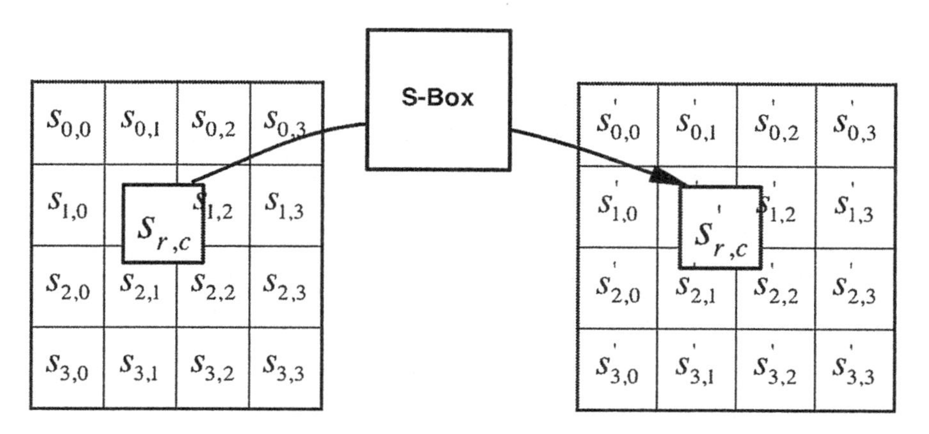

Fig. 8.7 The SubBytes substitution [2]

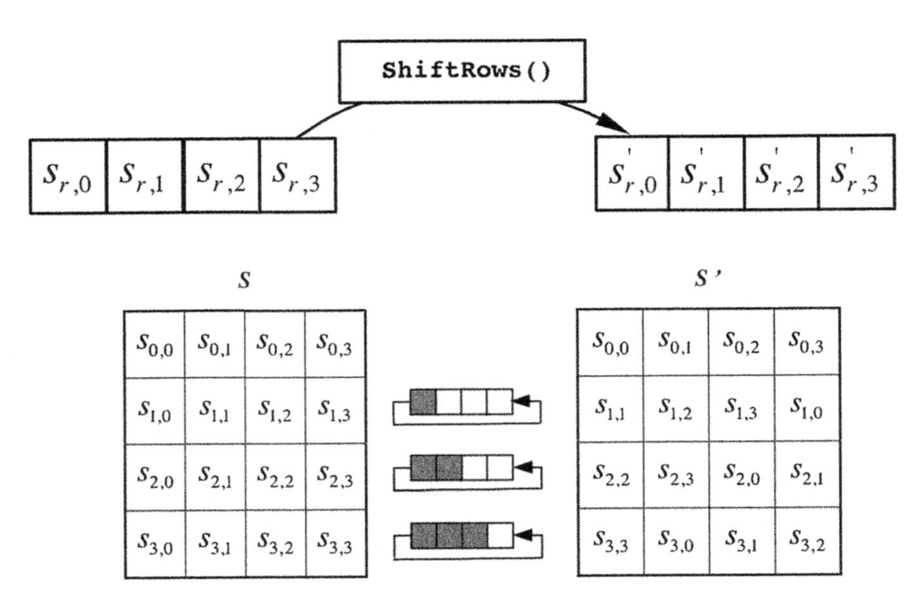

Fig. 8.8 ShiftRows function [2]

As opposed to DES, there has been no controversy with the adoption of Rijndael as the AES. This is because the entire process of picking the algorithm was open and transparent. After Rijndael was selected, the cryptographic community was given over a year to try to find flaws or weaknesses in the algorithm. The authors also published their own book on the design of the algorithm, providing their reasons for all their design decisions [4].

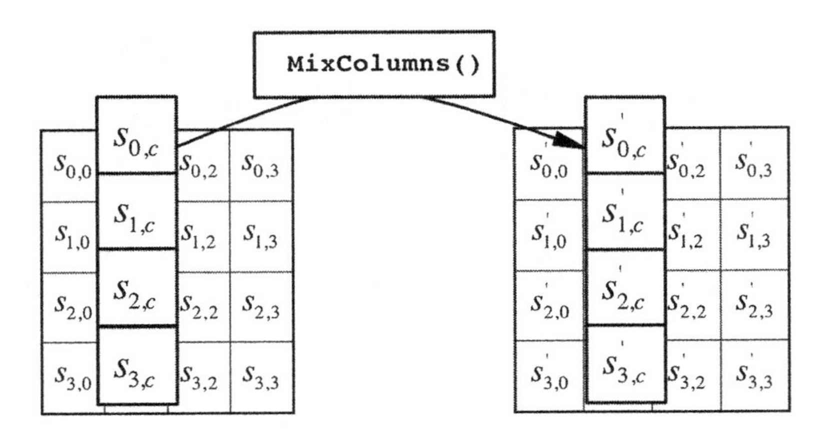

Fig. 8.9 The MixColumns function [2]

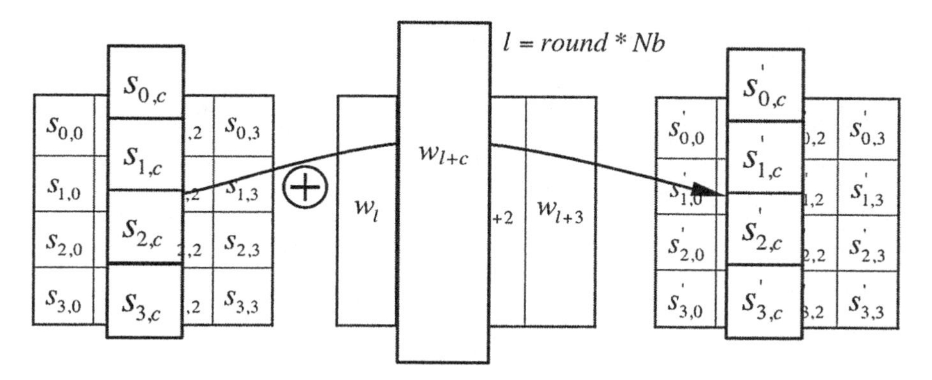

Fig. 8.10 The AddRoundKey function [2]

```
KeyExpansion(byte key[4*Nk], word w[Nb*(Nr+1)], Nk)
begin
   word  temp

   i = 0

   while (i < Nk)
      w[i] = word(key[4*i], key[4*i+1], key[4*i+2], key[4*i+3])
      i = i+1
   end while

   i = Nk

   while (i < Nb * (Nr+1)]
      temp = w[i-1]
      if (i mod Nk = 0)
         temp = SubWord(RotWord(temp)) xor Rcon[i/Nk]
      else if (Nk > 6 and i mod Nk = 4)
         temp = SubWord(temp)
      end if
      w[i] = w[i-Nk] xor temp
      i = i + 1
   end while
end
```

Fig. 8.11 The AES key scheduler [2]

References

1. Anonymous. 1999. Data Encryption Standard (DES) FIPS 46-3, ed. National Institute of Standards and Technology. Gaithersburg, MD: United States Department of Commerce.
2. Anonymous. 2001. Federal Information Processing Standard 197: Advanced encryption standard (AES), ed. National Institute of Standards and Technology. Gaithersburg, MD: United States Department of Commerce.

3. Bauer, Craig P. 2013. *Secret history: The story of cryptology*. Boca Raton, FL: CRC Press.
4. Daemen, Joan, and Vincent Rijmen. 2002. *The design of Rijndael: AES—The advanced encryption standard*. New York: Springer.
5. Diffie, Whitfield, and Martin Hellman. 1977. Exhaustive cryptanalysis of the NBS data encryption standard. *IEEE Computer* 10(6): 74–84.
6. Feistel, Horst. 1973. Cryptography and computer privacy. *Scientific American* 228(5): 15–23.
7. Hill, Lester S. 1929. Cryptography in an algebraic alphabet. *The American Mathematical Monthly* 36: 306–312.
8. Hill, Lester S. 1931. Concerning certain linear transformation apparatus of cryptography. *The American Mathematical Monthly* 38: 135–154.
9. Kahn, David. 1967. *The codebreakers: The story of secret writing*. New York: Macmillan (Hardcover).
10. Morris, R., N.J.A. Sloane, and A.D. Wyner. 1977. Assessment of the National Bureau of standards proposed federal data encryption standard. *Cryptologia* 1(3): 281–291.
11. Shannon, Claude. 1948. A mathematical theory of communication, parts I and II. *Bell System Technical Journal* 27: 379–423, 623–656.
12. Shannon, Claude. 1949. Communication Theory of Secrecy Systems. *Bell System Technical Journal* 28(4): 656–715.

Chapter 9
Alice and Bob and Whit and Martin: Public Key Crypto

Abstract The *key exchange* problem occurs with symmetric cipher systems because the same key is used for both enciphering and deciphering messages. This means that both the sender and receiver must have the same key and it must be distributed to them via a secure method. While this is merely inconvenient if there are only two correspondents, if there are tens or hundreds of people exchanging secret messages, then distributing keys is a major issue. Public-key cryptography eliminates this problem by mathematically breaking the key into two parts, a *public key* and a *private key*. The public key is published and available to anyone who wants to send a message and the private key is the only key that can successfully decipher a message enciphered with a particular public key. This chapter investigates the mechanisms used to implement public-key cryptography.

9.1 The Problem with Symmetric Ciphers

In his 1883 book *La Cryptographie Militaire*, the French cryptographer Auguste Kerckhoffs formulated a set of security rules for ciphers. The most important one is that you should always assume that the enemy knows the cipher system you are using. This implies that the entire security of the system must lie in the key. As long as the enemy doesn't have the key, they shouldn't be able to break the cipher.

For as long as *symmetric cipher* systems have been in existence—2,500 years or more—there has been a problem with using them. In order for a symmetric cipher to be used, both the sender and the receiver of a cryptogram must be in possession of the same key to unlock the cipher because that one key is used for both encryption and decryption. This means that everyone who is using a symmetric cipher system must have the same set of keys and they must use them in the correct order. We saw this in Chap. 6 when we observed that the Enigma day keys had to be distributed to all the users on an Enigma network every month so that

J. F. Dooley, *A Brief History of Cryptology and Cryptographic Algorithms*, SpringerBriefs in Computer Science, DOI: 10.1007/978-3-319-01628-3_9,

everyone would have the same day keys. The problem of synchronizing keys is known as the *key exchange* or *key distribution* problem.

With the advent of computer networks the need for more modern encryption systems to protect business data had grown. The NBS' call for algorithms and the imminent release of the Data Encryption Standard only made the key exchange problem more troublesome. Once the DES was released and people started making software products that used it, thousands or millions of keys would need to be exchanged.

Symmetric ciphers are used because there is an insecure communications channel (a telegraph, the postal system, a computer network) over which messages must be sent from one party to another. *Key distribution* for a symmetric cipher is done over a secure channel, usually using a courier, or by both parties meeting and exchanging the key. This is very inconvenient and time-consuming. Ideally, keys should be distributed over the same insecure channel as the encrypted messages. Of course to do this, one should encrypt the keys, but this requires a secure channel to share *that* key. This is the problem that needed to be solved for the last 2,500 years or so.

9.2 Enter Whit and Martin

In the early 1970s Whitfield Diffie and Martin Hellman were on opposite coasts but had the same problem. Both were interested in cryptology and both were thinking about the *key exchange* problem. Diffie had graduated from MIT with a degree in mathematics in 1965 and was an independent security consultant. Hellman had received his Ph.D. in electrical engineering from Stanford in 1969 and was now teaching there.

In 1974 Whitfield Diffie gave a talk at IBM's T.J. Watson Research Laboratory on key distribution. One of the audience members mentioned to him that a professor from Stanford University had also given a talk on key distribution a month or so before. That professor was Martin Hellman. Diffie jumped in his car and proceeded to drive cross-country to Palo Alto, California and visit Hellman. Hellman was dubious at first, but as the two talked, they discovered all the interests they had in common. Hellman agreed to take Diffie on as a graduate student and the two immediately began to work on the key exchange problem ([4, p. 256]).

As we will see, Diffie and Hellman were, in fact, working on three different problems.

9.3 The Key Exchange Problem

First was the traditional key exchange problem. How do you exchange a symmetric cipher key over an insecure channel? One way that Diffie and Hellman thought about this is with the "padlock" example ([4, p. 258, 1, p. 406]). Suppose that Alice and Bob (they are always called Alice and Bob) want to exchange messages with each other and

Eve wants to eavesdrop. For Alice to send a message she encrypts it using a key. For security purposes Alice uses a different key for each message. Bob must do the same thing. So you can see that Alice and Bob have to exchange many keys in order to keep up their correspondence. How do they do this? Well, one way to do this is for Alice to put her message in a box, and then lock the box with a padlock and key. Alice keeps the key and sends the box to Bob. But Bob doesn't have the key to Alice's padlock. So instead, Bob adds his own padlock to the box and sends it back to Alice. There are now two padlocks on the box, but now Alice can remove her padlock—she has the key to it—and send the box back to Bob, now with just Bob's padlock on it. Bob can now remove his padlock and then open the box to read the message. In this scenario Alice and Bob use two different keys and they don't have to share the keys with each other! But, the box goes back and forth several times, and the box has to allow either padlock to be added or removed in either order—the operation must be commutative.

This example can easily be applied to encryption, as long as one can encrypt a message twice and the order of encryption doesn't matter. We need to add a bit of notation to explain this. Let us assume that encryption and decryption are mathematical functions ([1, p. 406–407]). Let Alice's key be A, and Bob's key be B. Then $E_A(M)$ is the encryption function using Alice's key on message M. Similarly $E_B(M)$ is the encryption function using Bob's key on message M. $D_A()$ decrypts using Alice's key and $D_B()$ decrypts using Bob's key. So Alice first sends Bob $E_A(M)$. Bob then sends Alice $E_B(E_A(M))$. Then Alice sends Bob $D_A(E_B(E_A(M)))$ and Bob can finally do $D_B(D_A(E_B(E_A(M)))) = D_B(E_B(M)) = M$. Note that the functions D_A and E_B *must* commute or this scheme will not work.

So Diffie and Hellman had a proposed scheme to exchange secret keys over an insecure channel without the users having to meet or share another secret. Now all they needed was a cipher algorithm with the commutativity property. It turns out that traditional ciphers don't do this. Lets try an example. Lets say that Alice and Bob are using a monoalphabetic substitution cipher with a mixed alphabet as their encryption and decryption functions. They use different alphabets, but their functions need to commute. Here is what happens.

Alice's alphabet

```
a b c d e f g h i j k l m n o p q r s t u v w x y z
T V R S D B G M J Z E C L Q K U P X H Y I A O F W N
```

Bob's alphabet

```
a b c d e f g h i j k l m n o p q r s t u v w x y z
N J F K L Y M P O Q W A D U Z S I H C V E R B T G X
```

Message	m e e t	m e	a t	n o o n
$E_A(M)$	L D D Y	L D	T Y	Q K K Q
$E_B(E_A(M))$	A K K G	A K	V G	U Z Z U
$D_A(E_B(E_A(M)))$	V O O G	V O	B G	P J J P
$D_B(D_A(E_B(E_A(M))))$	t i i y	t o	w y	h b b h

The resulting output is gibberish because the substitution doesn't commute. But Diffie and Hellman were on the right trail; all they needed was a cipher algorithm—or a mathematical function that acted like one—that would commute properly. Eventually, after months of trying, Hellman found one and the two of them (with help from Ralph Merkle) fleshed it out.

The operation they chose is known as the *discrete log problem.* Say you choose a prime number p. Let another number g be a generator of the multiplicative cyclic group of integers *(mod p)*, called Z_p*. Then for values of x, the computation g^x (mod p) will generate all the elements of the group. The *discrete log* problem is given the group element g^a (mod p) what is a? It turns out that this is a very hard problem to solve, which is just the point. Here's how the Diffie-Hellman key exchange algorithm works.

First, Alice and Bob decide on the numbers g and p as described above and exchange them. They can do this over an insecure channel; it doesn't matter. Then, to exchange a secret key, Alice and Bob do the following. Each of them chooses a secret number, say Alice chooses a and Bob chooses b. These numbers they *must* keep secret. Alice then computes g^a (mod p) and Bob computes g^b (mod p). Both results are guaranteed to be elements of the group. Alice and Bob then exchange these new numbers; again this can be done over an insecure channel. Now, because the discrete log problem is hard, Alice can't really find b from the value g^b (mod p), and similarly Bob can't find a. But that's OK. If you remember the rules of exponentiation you'll remember that $(g^a)^b = (g^b)^a = g^{ab}$. Alice can now compute g^{ab} (mod p) and Bob can compute g^{ba} (mod p) and they both get a new number that is exactly the same, and that is now their secret key. As an example, suppose Alice and Bob decide that $p = 11$ and $g = 7$. So their one-way function is 7^x (mod 11). Now if Alice chooses $a = 2$ and Bob chooses $b = 6$ we have

$$7^2 \text{ (mod 11)} = 5 \text{ and } 7^6 \text{ (mod 11)} = 4$$

and Alice sends Bob a 5 and Bob sends Alice a 4. Alice and Bob now use these new numbers to compute the secret key. Alice now knows 7^6 (mod 11) $= 4$ and so she can now compute $(7^6)^2 = 4^2$ (mod 11) $= 5$. Bob now knows 7^2 (mod 11) $= 5$ and so computes $(7^2)^6 = 5^6$ (mod 11) $= 5$. Alice and Bob now each have an identical number that they can use as the key for any other cipher system they like.

We see that Diffie and Hellman have solved the first problem, key exchange. Using this system Alice and Bob can agree on a secret key for a symmetric cipher without needing to use a secure channel to exchange the key. The difficulty with this is that it requires several messages passing back and forth between Alice and Bob and it requires a semi-complicated mathematical operation. So this is where Diffie and Hellman move on to the second problem.

9.4 Public-Key Cryptography Appears

The next problem is how do Alice and Bob communicate without exchanging several messages every time they want to change keys. Ideally they would need to exchange nothing or at most one piece of information in order to communicate securely. How is this accomplished?

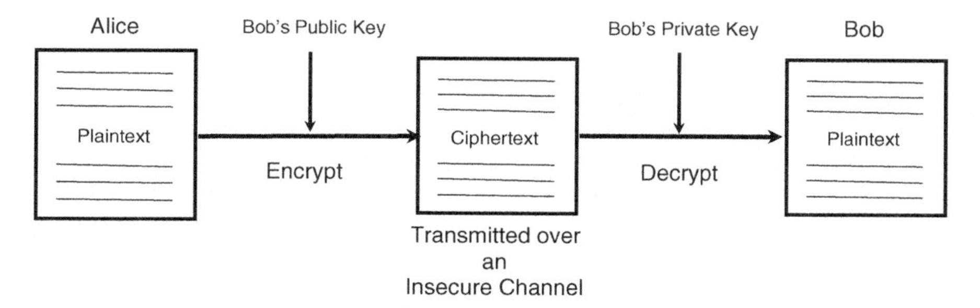

Fig. 9.1 The basic model of public-key cryptography

This time it was Diffie who had the initial breakthrough. In a symmetric cipher, the same key is used for encryption and decryption. Why not separate these operations and *use different keys* for encryption and decryption instead? In other words, make the system *asymmetric*. Figure 9.1 is a diagram of what Diffie and Hellman had in mind.

In this *asymmetric cipher* model, Alice and Bob each have two keys, a *public key* that anyone can see and is used to encrypt messages to that person, and a *private key* that is kept secret and is used to decrypt messages. So if Alice wants to send a message to Bob, she gets Bob's public key—it can be published in a key directory—and encrypts her message using Bob's public key. She then sends the message to Bob. When Bob receives the message he uses his private key, which only he knows, to decrypt Alice's message. Even if Eve (remember Eve?) intercepts Alice's message she can't decrypt it because only the person with Bob's private key can decrypt a message enciphered with his public key and Bob keeps his private key secret from everyone.

So finally we have a cipher system that does not require any key exchange at all. Everyone who wants to communicate generates a public–private key pair and publishes their public key in some kind of directory. Anyone who wants to send them a message grabs their public key, enciphers the message, and ships it off. Decrypting uses the secret private key which is never transmitted anywhere to anyone. This system is called *public-key cryptography*.

Only two small problems remain. What algorithm to use to generate the public–private key pairs and how to do encryption and decryption? Alas, in their paper [2] that describes both the *key exchange* solution and *public-key cryptography*, Diffie and Hellman don't provide an algorithm. That remained for others to do.

9.5 Authentication is a Problem Too

The third problem that Diffie and Hellman tackle in their foundational paper on public-key cryptography is that of authentication. If Bob receives an encrypted message from Alice how can he be sure that Alice really sent that message? After all, anybody can use Bob's public key to encrypt a message and send it. Just appending Alice's signature to the bottom of the message is not really proof that the message is from Alice. So how can Bob be sure that Alice really sent the message with her name on it?

It turns out the solution to authentication, also called a *digital signature*, is right in the public key algorithm itself. One of the requirements of the public–private key pair is that they must be able to be applied to the message in any order; they must commute. Given this requirement, the process illustrated in Fig. 9.2 will accomplish authentication.

Say that Alice wants to send a message to Bob and she wants to guarantee that Bob knows it is from her. Using the Digital Signature process she will first encrypt the message with Bob's public key, ensuring that only Bob can read it. Then, she encrypts the encrypted message with her private key and sends it off to Bob. When Bob receives Alice's message he decrypts it first using Alice's public key (this is OK because the public–private key pairs commute), revealing the inner message that he then decrypts using his own private key. If the resulting plaintext is a readable message Bob is then assured that Alice must have sent the message because she is the only person who has the private key that matches her public key. The digital signature process thus provides the necessary authentication protocol for public-key cryptography.

9.6 Implementing Public-Key Cryptography: The RSA Algorithm

The publication of Diffie and Hellman's *New Directions in Cryptography* in November 1976 [2] was a landmark in computer cryptography. It also started a race to see who could come up with the algorithms necessary to implement the public–private key generation and the encryption and decryption algorithm itself.

The first publication that met all the requirements of the system was authored in 1977 by three professors at MIT, Ronald Rivest, Adi Shamir, and Leonard Adleman [3]. The RSA algorithm is based on exponentiation in a finite (Galois) field over integers (mod p) where p is a prime. Here is how the algorithm works:

First Alice chooses two large prime numbers p and q. She will keep these numbers secret. She then computes their product $n = p * q$.

Next she computes the *Euler totient function* for n $\varphi(n) = (p-1)(q-1)$. This is the number of numbers less than n and relatively prime to n (they have no common factors). For a prime number p there are always $(p-1)$ numbers relatively prime to it. Alice then selects a number e where $1 < e < \varphi(n)$ and such that $\gcd(e, \varphi(n)) = 1$, that is e and $\varphi(n)$ must be relatively prime as well. Alice can tell if e and $\varphi(n)$ are relatively prime by using the *Euclidean algorithm* to compute the greatest common divisor of the two numbers, $\gcd(e, \varphi(n)) = 1$. Since e is relatively prime to $\varphi(n)$ then it must have a multiplicative inverse (mod $\varphi(n)$), called d. This means that $e * d = 1 \pmod{\varphi(n)}$.

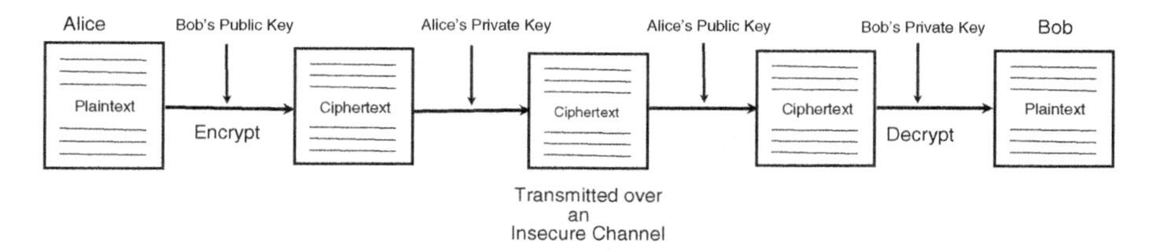

Fig. 9.2 The digital signature process

Alice must next compute d. This can be done easily using a variation on the Euclidean algorithm called the *Extended Euclidean algorithm.* Alice is now ready to proceed. She has a pair (e, n) that is her public key, and she has a pair (d, n) that is her private key.

Alice now publishes her public key pair in a directory that is available to anyone who wants to send her a secret message.

To encrypt a message in RSA, Bob must first convert the message M into a number. This is usually easy to do because, after all, underneath the hood all letters are represented by numbers on a computer. Bob may need to break his message M up into several parts in order to convert it to a number. Once he has converted his message Bob is ready to encrypt.

Next Bob retrieves Alice's public key (e, n) from the directory.

He computes $C = M^e \pmod{n}$. That is, he raises his message to the power of e and then reduces the product (mod n). The result is his ciphertext.

Bob then sends this off to Alice.

To decrypt Bob's message, Alice must use her private key (d, n).

She computes $M = C^d \pmod{n}$ to retrieve the message that Bob sent. This computation retrieves the message because e and d are multiplicative inverses (mod n). To clarify this process we'll do an example.

9.6.1 RSA Key Generation

Select two primes, $p = 17, q = 11$.

Compute $n = p * q = 17 * 11 = 187$

Compute $\varphi(n) = (p-1) * (q-1) = 16 * 10 = 160$

Select e such that $\gcd(e, 160) = 1$. We'll choose $e = 7$

Determine d: $e * d = 1 \pmod{160}$ and $d < 160$. For us $d = 23$ because $23 * 7 = 161 = 1 \pmod{160}$.

Publish the public key (7, 187) and

keep secret the private key (23, 187)

9.6.2 Encrypting and Decrypting

Now to encrypt we need a message. Say M = 88

$C = 88^7 \pmod{187} = 11$ This is our ciphertext.

Now to decrypt we take the ciphertext and undo the encryption

$M = 11^{23} \pmod{187} = 88$

9.7 Analysis of RSA

The security of the RSA algorithm lies in two areas. First, while it is easy to compute n = p * q, it is very difficult to do the reverse. That is, it is extremely computationally expensive to find the *prime factors* of a large composite number. This is the lynchpin of RSA security.

That leads us to the other part of the security of RSA. The two prime numbers p and q must be very large primes; large enough so that their binary representations convert to around 500 bits or more each. This will lead to a binary product of around 1,000 bits. As computers get faster and faster, the number of bits in n = p * q will need to grow. So how big should your keys be for RSA? RSA Laboratories, the company founded by the three authors suggests:

> RSA Laboratories currently recommends key sizes of 1,024 bits for corporate use and 2,048 bits for extremely valuable keys like the root key pair used by a certifying authority (see Question 4.1.3.12). Several recent standards specify a 1,024-bit minimum for corporate use. Less valuable information may well be encrypted using a 768-bit key, as such a key is still beyond the reach of all known key breaking algorithms [From http://www.rsa.com/rsalabs/node.asp?id=2218. Retrieved on 06/21/2013].

Cipher systems that implement public-key cryptography have roughly the same security as equivalent symmetric key systems with key lengths about one-third the length of the RSA key. So why haven't public key systems replaced symmetric systems over the last forty years or so? The answer is speed. It turns out the public key systems are slow, in some cases very slow. Symmetric systems like DES and AES use very simple computer operations like exclusive or and bit shifting. To date all the public key systems developed require complicated mathematical operations to work. These mathematical functions require much more CPU time than the simple operations required for symmetric systems. This has limited public key systems primarily to the role for which they were first envisioned—solving the key exchange problem.

9.8 Applications of Public-Key Cryptography

The most frequent application of public-key cryptography is in Internet commerce. Every time you make a transaction with Amazon you are using the RSA algorithm and public-key cryptography. The RSA algorithm is used to encrypt a symmetric key and send it from the client to the server and that key is then used to handle all the encryption of the rest of your transaction with the server. Here's how it works.

Your browser implements a communications protocol called SSL/TLS (Secure Socket Layer/Transport Layer Security). That communications protocol is set up by establishing a common cryptographic algorithm between the client (your computer) and the server (the web host you are talking to). A handshaking protocol is used to establish communications and transfer the key to the symmetric cipher system used by the two machines. Here is how the system establishes the link to the server:

1. The *client* sends the server the client's SSL version number, cipher settings, session-specific data, and other information that the server needs to communicate with the client using SSL.
2. The *server* sends the client the server's SSL version number, cipher settings, session-specific data, and other information that the client needs to communicate with the server over SSL. The server also sends its own certificate, and if the client is requesting a server resource that requires client authentication, the server requests the client's certificate. The certificate includes the servers RSA public key.
3. The client uses the information sent by the server to authenticate the server. If the server cannot be authenticated, the user is warned of the problem and informed that an encrypted and authenticated connection cannot be established. If the server can be successfully authenticated, the client proceeds to the next step.
4. Using all data generated in the handshake thus far, the client (with the cooperation of the server, depending on the cipher in use) creates the pre-master secret key for the session, encrypts it with the server's public key, and then sends the encrypted pre-master secret key to the server.
5. If the server has requested client authentication (an optional step in the handshake), the client also signs another piece of data that is unique to this handshake and known by both the client and server using it's RSA private key. In this case, the client sends both the signed data and the client's own certificate containing the client's RSA public key to the server along with the encrypted pre-master secret.
6. If the server has requested client authentication, the server attempts to authenticate the client. If the client cannot be authenticated, the session ends. If the client can be successfully authenticated, the server uses its RSA private key to decrypt the pre-master secret key, and then performs a series of steps (which the client also performs, starting from the same pre-master secret) to generate the master secret key.
7. Both the client and the server use the master secret key to generate the session key, which is a symmetric key used to encrypt and decrypt information exchanged during the SSL session and to verify its integrity (that is, to detect any changes in the data between the time it was sent and the time it is received over the SSL connection).
8. The client sends a message to the server informing it that future messages from the client will be encrypted with the session key. It then sends a separate (encrypted) message indicating that the client portion of the handshake is finished.

9. The server sends a message to the client informing it that future messages from the server will be encrypted with the session key. It then sends a separate (encrypted) message indicating that the server portion of the handshake is finished.

[From http://en.wikipedia.org/wiki/Transport_Layer_Security retrieved 06/21/2013].

Every time you buy a book from Amazon, you are using the RSA algorithm to transfer symmetric keys that are used to finish passing the data of your transaction back and forth between your computer and the Amazon web server.

References

1. Bauer, Craig P. 2013. *Secret history: The story of cryptology*. Boca Raton, FL: CRC Press.
2. Diffie, Whitfield, and Martin. Hellman. 1976. New directions in cryptography. *IEEE Transactions on Information Theory* IT 22(6): 644–654.
3. Rivest, R.L., A. Shamir, and L. Adleman. 1978. A method for obtaining digital signatures and public key cryptosystems. *Communications of the ACM* 21(2): 120–126.
4. Singh, Simon. 1999. *The code book: The evolution of secrecy from Mary, Queen of Scots to quantum cryptography*. New York, NY: Doubleday.

Index

J. F. Dooley, *A Brief History of Cryptology and Cryptographic Algorithms*,
SpringerBriefs in Computer Science, DOI: 10.1007/978-3-319-01628-3,

CPSIA information can be obtained at www.ICGtesting.com
Printed in the USA
LVOW04s0410111214

418277LV00014B/540/P

9 783319 016276